# **Die Schwingen des Kranichs** 50 Jahre Lufthansa Design
# **The wings of the crane** 50 years of Lufthansa design

Lufthansa
D-ABVA

Volker Fischer

# Die Schwingen des Kranichs
50 Jahre Lufthansa Design

# The wings of the crane
50 years of Lufthansa design

Edition Axel Menges

ISBN 3-932565-53-3



Das vorliegende Buch begleitet die gleichnamige Ausstellung im Museum für Angewandte Kunst in Frankfurt am Main vom 16. 6. bis 4. 9. 2005.
The present book appears in conjunction with the exhibit of the same name, being held at the Museum for Applied Arts in Frankfurt am Main from June 16 to September 4, 2005.

Frontispiz: Boeing 747-400, ab Ende der 1980er Jahre
Frontispiece: Boeing 747-400, as of the late 1980s

**Buch**
**Book**

**Konzept und Text**
**Concept and copy**
Volker Fischer

**Recherche**
**Research**
Volker Fischer, Christine Ramme

**Übersetzung ins Englische**
**Translation into English**
Stewart Lindemann

**Übersetzung aus dem Englischen**
**Translation from English**
Volker Fischer

**Fotografie und Scans**
**Photography and scans**
Christine Ramme; Friederike Biegel, Uta Kammerer, Peter Schmidt Group, Frankfurt am Main

**Gestaltung**
**Design**
Peter Schoemig, Peter Schmidt Group, Hamburg

**Koordination**
**Coordination**
Ralf-Dieter Rudorf, Eenita Struve, Deutsche Lufthansa AG; Friederike Biegel, Peter Schmidt Group, Frankfurt am Main

**Druck**
**Printing**
Druckhaus Münster GmbH, Kornwestheim

**Bindearbeiten**
**Binding**
Buchwerk GmbH, Darmstadt

**Ausstellung**
**Exhibit**

**Konzept**
**Concept**
Volker Fischer; Georg Schubert, Deutsche Lufthansa AG

**Leihverkehr und Ausstellungsdesign**
**Organisation of loans and exhibit design**
Volker Fischer; Anja Naumann, Daniel Mertes, GPGO events & messen GmbH; Christine Ramme

**Grafik Medien**
**Media graphics**
Peter Schoemig, Peter Schmidt Group, Hamburg

**Grafik Ausstellung**
**Exhibit graphics**
vier5, Paris; Antje Riemann

**Presse- und Öffentlichkeitsarbeit**
**Public relations**
Yvonne von Kienlin; Amelie Lorenz, Deutsche Lufthansa AG

Allen Leihgebern danken wir für ihre freundliche Unterstützung.
We want to thank all those who have loaned materials for their kind support.

# Inhalt
# Content

# Design, das beflügelt
# Design that lends wings

Liebe Leserinnen und Leser,

2005 ist ein bedeutendes Jahr für Lufthansa: Ein großer Name, eine starke Marke wird 50. Aus einer kleinen deutschen Fluggesellschaft in öffentlicher Hand ist ein vollständig privatisierter, erfolgreicher und international aufgestellter Luftfahrtkonzern geworden. Unverändert blieb unser Markenversprechen, auch wenn der Wandel im Erscheinungsbild der Lufthansa erkennbar ist. Der Kranich, unser Markenzeichen, hat im Laufe der Jahre an Strahlkraft gewonnen.

Wir haben in den letzten 50 Jahren immer wieder internationale Standards gesetzt. Lufthansa steht für Sicherheit und Zuverlässigkeit, für Leistungsfähigkeit und Innovationskraft, für Qualität und Service, für Kompetenz und Klarheit. Das drückt sich auch in unserem Design aus. Die besten Designer ihrer Zeit haben das Erscheinungsbild der Lufthansa kreiert und später kontinuierlich weiterentwickelt. Es ist präsent: Unsere Verkaufsbüros, die Ticketschalter, die Lounges, das Bordgeschirr, die Dienstuniformen unserer Mitarbeiter oder die Flugzeuge – das Design von Lufthansa und dessen Veränderungen sind sichtbar. Lufthansa ist heute mit ihren 50 Jahren jünger, frischer, bunter und erfolgreicher denn je.

Sie werden in diesem Buch vieles wiedererkennen, wenn Sie in die Zeit eintauchen, die Sie entweder selbst noch erlebt haben oder aus Erzählungen kennen. Dabei wünsche ich Ihnen viel Freude. Seien Sie versichert, dass auch der Designgeschichte der Lufthansa noch viele Erfolgskapitel hinzugefügt werden. Wir freuen uns darauf.

Wolfgang Mayrhuber
Vorsitzender des Vorstands
Deutsche Lufthansa AG

Dear Readers,

2005 is a milestone year for Lufthansa: A great name and a strong brand are turning 50. What began as a small, government-owned German airline has become a fully privatized and highly successful air carrier of international renown. Unchanged is the promise behind our brand, this in spite of modifications in the Lufthansa corporate image. The soaring crane, our brand emblem, has gained in radiance over the years.

In the past 50 years we have set international standards again and again. Lufthansa symbolizes safety and reliability, performance and innovation, quality and service, expertise and clarity. All these attributes find expression in our corporate design. The finest designers of their time were engaged to shape the Lufthansa corporate identity and later to continuously develop and refine that identity. It is present everywhere you look: in our sales offices, at ticket counters, in the lounges. It is reflected in the airline dinnerware, in our employees' uniforms and in the aircraft themselves. The Lufthansa design and its changes are quite apparent. Today, at 50 years of age, Lufthansa is younger, fresher, more colorful and more successful than ever before.

In this book you will surely recognize many things as you dip into the past, things which you might have experienced yourself or which you have heard or read about. I hope you will enjoy this journey through time. You may be assured that we will be adding many success chapters to Lufthansa's illustrious design story in the future. We are looking forward to the challenges of change.

Wolfgang Mayrhuber
Chairman of the Executive Board
and CEO Deutsche Lufthansa AG

LH 574

Ikarus, ca. 1500 v. Chr.

Lilienthal, 1891

Gebrüder Wright, 1903

Montgolfiere, 1783

Junkers Ju 52, 1930

Lufthansa, 1974

# Heute gehört zum Fliegen noch ein bißchen mehr als Idealismus. Der Wille zur Perfektion.

**Lufthansa**

Flugscheine erhalten Sie in über 700 IATA-Reisebüros mit Lufthansa-Agentur.

# Beflügelt

Seit der griechischen Sage von Daidalos und seinem Sohn Ikarus ist Fliegen ein archetypischer Traum des Menschen. Er ist die Bedingung der Möglichkeit, „Fliegen-Können" überhaupt für denkbar zu halten und insofern auch die unmittelbare Bedingung für alle technischen Realisierungen dieser Art – von Leonardos Flugmaschine über das erste Laufflugzeug der Gebrüder Wright bis zu den großen Passagierflugzeugen unserer Tage wie die Boeing- und Airbus-Maschinen, die auch die Lufthansa fliegt.

„vermutlich jeder mensch ist im traum einmal selbst geflogen. meist bekommt er schwierigkeiten, sobald er das gefühl hat, wirklich zu fliegen. wird er weiterhin genügend auftrieb haben, genügend balance, wird er abstürzen? fliegen ist nicht nur ein traum der menschheit, fliegen ist ein traum der natur, des lebens selber. die entwicklung der lebewesen beginnt im wasser. wasser hat tragkraft und verschafft geborgenheit. aber das leben hat mechanismen entwickelt, um auch aufs land zu kommen, und es hat bis in unsere träume hinein demonstriert, fliegen zu wollen. das leben hat zahllose versuche unternommen, sich in die lüfte zu erheben. mit erfolg. was sicherlich auch die träume des menschen beflügelt hat. in unserem jahrhundert hat es auch der mensch geschafft, sich in die lüfte zu erheben, nicht etwa mit seiner physis, sondern mit hilfe von fluggeräten ..."[1]

Eine Geschichte der designrelevanten Formgestaltung, des visuellen Erscheinungsbildes und der werblichen Selbstdarstellung eines Unternehmens wie der Lufthansa ist ein disziplinäres Unterfangen, welche manche, möglicherweise für ein Gesamtbild ebenso wichtige Aspekte und Facetten vernachlässigt. Die wirtschaftliche Entwicklung, die Personalstruktur, die Bilanzen, die internen Finanz- und Beteiligungsstrategien, die Gebäude und Grundstücke ebenso wie die technische Entwicklung der Flugzeuge und der Bodenfahrzeuge muss einer allgemeineren Darstellung vorbehalten bleiben. Im folgenden geht es ausschließlich um jene Zusammenhänge der Lufthansa, die gestalterische Entscheidungen charakterisieren: um ihr Erscheinungsbild, ihre Counter, Lounges und Stadtbüros am Boden, die Bordausstattung der Maschinen, um die Tischkultur an Bord, um die Uniformen des Personals, um die Warenangebote an Bord, um die Werbung der Lufthansa und um die visuelle Erinnerungskultur in persönlichen Memorabilien. Aber gerade im Progress dieser gestalterischen Entscheidungen spiegelt sich auch die Kontur und Kultur des gesamten Unternehmens nach Innen und Außen. Eine solche ästhetische Annäherung an ein üblicherweise dominant technisch gesehenes Phänomen wie eine Fluggesellschaft kann spannende Einsichten vermitteln. Denn in nur 80 Jahren ist es der zivilen Luftfahrt gelungen, eine ganz eigene Ästhetik zu entwickeln: es entstand gewissermaßen ein ganz neues, unverwechselbares Design-Universum, an dem die Lufthansa von Anfang an beteiligt war. 1926 als Zusammenschluss der Deutschen Aeroloyd und von Junkers-Luftverkehr gegründet, wurde sie in der Folge des Zweiten Weltkrieges zunächst von den Alliierten liquidiert und erst 1955 neu etabliert. Am 1. April dieses Jahres begann erneut der innerdeutsche Luftverkehr und noch im gleichen Jahr wurden zunächst Madrid, London und Paris, dann auch schon New York angeflogen. Niemand konnte sich ausmalen, dass die Lufthansa bereits ein gutes Jahrzehnt später zu den führenden Fluggesellschaften der Welt zählen würde. Am Anfang flogen noch amerikanische Piloten die LH-Maschinen, denn deutsche Piloten mussten erst ausgebildet werden. In jedem Folgejahr verdoppelten sich die Anflughäfen und auch das Betriebskapital. Kein Zweifel, die Lufthansa war von Anfang an

Seite 8: Anzeige, 1974
Page 8: Advertisement, 1974

# A high-flying story

Ever since the days of the Greek saga telling the story of Daedalus and his son Icarus, flying has been an archetypal dream of humankind. That dream is the precondition for thinking that flying is at all possible and thus the direct precondition for every technical advancement toward this goal – for everything from Leonardo's flying machine and the Wright brothers' first flight at Kitty Hawk to the giant passenger airliners of our days such as the Boeing and Airbus aircraft flown by Lufthansa.

"Presumably every person has at one time flown in the course of a dream. Usually the dreamer encounters difficulties as soon as he has the feeling that he is actually flying. Will he have enough lift, stay in balance; will he crash? Flying is not only a dream of humans, it is a dream of nature and of life itself. The development of living creatures began in the water. Water has carrying power and creates a feeling of refuge. But life developed mechanisms so that it could make its way on land and has demonstrated, right into our dreams, that it wants to fly. Life has made numerous attempts to lift itself into the air. With success. And that has certainly given wing to men's dreams. In our century humans also have managed to rise into the air, not with their own physique but with the help of flying machines ..."[1]

A history of design, of the visual identity and the promotional representation of a company like Lufthansa is a mono-disciplinary undertaking which might neglect aspects and facets which are just as important in seeing the overall pictures. The economic development, personnel structure, balance sheets, internal strategies for financing and holdings, the buildings and properties, as well as the technical development of the aircraft and the ground vehicles will have to be reserved to a broader depiction of the corporation. In the following chapters we will deal exclusively with those Lufthansa contexts which are affected by the design decisions: the corporate design, the counters, lounges and downtown offices on the ground, the cabin interiors, inflight dining culture, the uniforms worn by the employees, the products offered in flight and on the ground, the Lufthansa advertising efforts and the visual culture of recollection in the form of personal memorabilia. But it is exactly the progress of these design decisions which reflects the contours and culture of the entire company, both inwardly and outwardly. This esthetic approach to a phenomenon such as an airline, which is otherwise seen predominantly as a technical matter, can provide exciting insights. That is because in just 80 years civil aviation has been able to develop its own, very distinctive esthetics canon. Created here was, so to speak, an entirely new and unmistakable design universe, a development in which the Lufthansa participated from the very beginning. Founded in 1926 with the merger of Deutsche Aero Lloyd and Junkers-Luftverkehr, it was first disbanded by the Allies after World War II and was not reestablished until 1955. On April 1 of that year domestic air travel recommenced and in the very same year Madrid, London and Paris – and even New York City – were added as destinations. No one could have imagined at the time that, just a good decade later, Lufthansa would already have become one of the world's leading air carriers. At the beginning American pilots flew the Lufthansa airliners since German pilots had yet to be trained. In each year which followed the number of destinations doubled and with that the operating capital. There is no doubt that Lufthansa was on a success path from the very beginning. Today Deutsche Lufthansa transports more than 120,000 passengers every day, and thus well over

Image-Anzeige, 1974
Image advertisement, 1974

auf Erfolgskurs. Heute befördert die Deutsche Lufthansa täglich über 120.000 Fluggäste, also deutlich mehr als dreieinhalb Millionen Passagiere pro Monat. Alle 6 Sekunden startet eine LH-Maschine.

Alle Fluggesellschaften sind zunächst Verkehrsunternehmen, die ihre Passagiere möglichst zeiteffizient von A nach B bringen. Im Unterschied aber zu Ozeandampfern oder Eisenbahnen sind bei Fluggesellschaften die Distanzen weiter, obwohl inzwischen der Passagierpreis erheblich günstiger ist. Das subjektive Gefühl der Reisegefahr ist hoch, obwohl Fliegen statistisch die sicherste Art des Reisens ist. „Die ambivalente seelische Erschütterung zwischen Euphorie und Panik, die die Überwindung der Schwerkraft auslöst, ist freilich bis heute nicht gänzlich ‚verflogen'."[2] Aus diesem metaphysischen Unbehagen, welches subjektiv, aber eben nicht objektiv begründbar ist, erklärt sich der besondere visuelle Diskurs all jener Gegenstände, die als Schnittstellen das Flugerlebnis der Passagiere begleiten und kulturalisieren. Die Flugzeuginterieurs, das Corporate Design der Gesellschaften, die Uniformen der Flugbegleiterinnen, das Bordgeschirr, die Freundlichkeit des visuellen Erscheinungsbildes, die Ansprache über Werbung, die Unterhaltungselektronik ebenso wie das Essen und Trinken an Bord sind insgesamt Mittel und Medien, Gegenstände und Gadgets, die vom Vorgang des Fliegens eher ablenken sollen und sich auf Erlebnisbereiche „am Boden" beziehen:

- Die Flugsitze, auf Langstrecken inzwischen zu komfortablen Betten umzubauen, bieten sowohl die Entspannung eines geruhsamen Nickerchens als auch den Komfort eines „Home-Entertainments" durch integrierte Audio bzw. TV-Unterhaltung.
- Das Bordgeschirr, die Speisekarten wie überhaupt das Dinieren an Bord partizipieren am Bedeutungssystem der gehobenen Gastronomie und immer mehr an jenem Kontext, der heute „Erlebnisgastronomie" genannt wird.
- Die Fluglinien-Werbung spricht wahlweise touristische Abenteuer, Exotismus, Relaxen, Urlaub vom Alltag, fremde Länder und fremde Sitten an. Auch wirbt sie regelmäßig mit der technischen Brillanz der Flugzeuge, dem Know how ihrer Techniker und der Freundlichkeit des Personals am Boden und an Bord.
- Über Jahrzehnte hinweg war es, zumindest auf längeren Flügen, üblich, den Passagierraum auf Zeit mit großformatigen Vorführungen attraktiver Spielfilme zum Kinosaal umzurüsten, die über Kopfhörer in verschiedenen Sprachen erlebbar waren. Heute haben miniaturisierte Vidoesysteme in den Rücken- oder Armlehnen der Flugsitze die bewegte Bildunterhaltung übernommen und nochmals individualisiert, weil Dutzende verschiedener Filme „on demand" zur Verfügung stehen.

Bis heute ist ein großer Teil des Bedeutungsraumes der mit dem Fliegen verbundenen Begriffe und Funktionen maritim kodiert: Kapitän, Erster bzw. Zweiter Offizier, Steward, Maschinenflotte, Passage, Passagier, an Bord bzw. von Bord gehen, Bordverpflegung, Senatorklasse etc. Auch der Name Lufthansa – vor dem Krieg auseinander geschrieben und damit noch deutlicher ein Hinweis auf seine Herkunft – lässt ja nicht nur den Begriff der „Hanse", jene genossenschaftlichen Vereinigungen besonders norddeutscher Kaufleute im 12. und 13. Jahrhundert, die sich auch im Ausland gegenseitig unterstützten, sondern auch den Begriff der Luftschifffahrt assoziieren, der nochmals deutlich auf das Meer verweist. Im Übrigen trugen am Beginn der Lufthansa sowohl das Boden- wie das fliegende Personal tatsächlich Schiffsuniformen. Beides, zur See fahren und fliegen, setzt sich den Elementen aus, dem Wasser und der Luft, den Strömungen, Winden, Turbulenzen und Stürmen. Selbst das Segeln in der Luft ist sprachlich vom Segeln auf dem Wasser abgeleitet. Und es mutet fast wie ein Treppenwitz der Linguistik an, dass in jüngerer Zeit der schienengebundene Verkehr à la ICE dieses Vokabular vom Flugbetrieb zurückübernommen hat: da begrüßt via Lautspre-

three and a half million per month. A Lufthansa airplane takes off every six seconds.

Every airline is in the main a transportation undertaking which moves its passengers from point A to point B as quickly as possible. When compared with the ocean liners or railways, the distances which the airlines cover are far greater, even though the price for passage has become considerably more affordable. The subjective feeling of hazard is high even though statistics show that flying is the safest way to travel. "The ambivalent mental vacillation between euphoria and panic which is triggered by overcoming gravity has not been entirely subdued."[2] This metaphysical discomfort, which can be justified subjectively, but by no means objectively, explains the special visual discourse among all those objects which, as interfaces with aviation itself, accompany and "culturalize" the passenger's experience in the air. The aircraft interiors, the company's corporate design, the flight attendants' uniforms, the tableware, the friendliness of the visual identity, the approach taken by advertising, the entertainment electronics and eating and drinking on board are all media, objects and gadgets which are intended to distract attention from the act of flying and make reference to realms of experience found "on the ground":

- The airline seats, which in some cases can be converted into comfortable beds for long-haul flights, offer both the relaxation of a quiet nap and the comfort of "home entertainment" using integrated audio and TV components.
- The airline dinnerware, the menus and dining on board all make use of the system of values found in first-rate restaurants and ever more often draw upon the context which is today referred to as "experience dining."
- Airline advertising mentions alternately tourist adventure, exotic places, relaxation, and an escape from the everyday world, to foreign countries and foreign customs. It also regularly emphasizes the technical brilliance of the aircraft, the expertise of the technicians and the friendliness of the personnel, both on the ground and in the air.
- For decades standard procedure, at least on longer flights, was to temporarily convert the passenger cabin into a cinema with large-screen showings of popular movies, the sound track being delivered in several languages over headphones. Today miniature video systems in the backrests or armrests provide "motion-picture entertainment" and have individualized it once again, since dozens of different movies are available "on demand."

Down to the present day a large part of the associations for the concepts and functions connected with flying have a maritime background: captain, first or second officer, steward, fleet, passage, boarding, disembarking, on-board dining, Senator Class etc. The name Lufthansa was written as two words – Luft Hansa – before the War and thus made even clearer reference to its heritage. The first term is simply the German word for "air." The second harks back to the term "Hanse" – a cooperative association of merchants, particularly those located in northern Germany in the twelfth and thirteenth centuries, who provided mutual support in foreign fields. Implied is the concept of "airships," which once again makes a clear reference to the seas. At the beginnings of the Lufthansa both ground and air crews actually wore maritime uniforms. Both these experiences – setting out to sea and climbing into the air – expose the traveler to the elements: water and air, currents, winds, turbulence and storms. Even airborne "sailing" – in sailplanes – is derived linguistically from sailing on the water. And it seems almost like a paradox of history that rail transport on new high-speed ICE trains has recently adopted the vocabulary of flight operations. There an employee greets

chersystem ein Mitarbeiter die Fahrgäste „an Bord" und er heißt eben nicht mehr nur einfach „Schaffner", sondern in Anlehnung an den Begriff des Flugbegleiters nun Zugbegleiter; das Restaurant nennt sich „Bordrestaurant" und der Lokführer spricht via Mikrofon zu den Fahrgästen als „Ihr ICE-Kapitän". Unbenommen davon sind sowieso im Flugbetrieb der Anglizismen gar viele: Trolley, Purser, Counter, Skytrain als aufgeständerte Zubringerhochbahn, Boarding, Catering usw., zu schweigen vom anglophilen Speziolekt der Flugkapitäne via Tower und Fluglotsen.

All diese Erfahrungssplitter der Ansprache der Fluggäste durch das Flugunternehmen verweisen auf einen für das Fliegen eigentümlichen Referenzraum, der Fortschritt und Nostalgie, Internationalität und Regionalismus zu mischen weiß. Obwohl Fluggesellschaften nationengeprägt sind, agieren sie immer transnational und schon aus wirtschaftlichen Gründen kulturkreisübergreifend: so ist etwa die Star Alliance das erste wirklich weltweite Bündnis von Fluggesellschaften, welches 1997 von den fünf Gründungsmitgliedern Air Canada, SAS, Thai Airways International, United Airlines und Lufthansa gegründet wurde. Im Laufe des gleichen Jahres traten diesem Bündnis dann die Fluggesellschaften Varig, Ansett Australia, Air New Zealand sowie All Nippon Airways bei. Ebenso wie das Bonusprogramm Miles & More ist auch die Star Alliance mit mittlerweile über 15 Fluggesellschaften ein im globalen Maßstab konzipiertes „Mehrwert"-System.

Wenn wir im folgenden den Blick auf die gestalterischen Entscheidungen der letzten fünf Jahrzehnte der Lufthansa richten, wird deutlich werden, dass sich das Fliegen vom exklusiven Vergnügen betuchter Passagiere zum alltäglichen Massenverkehrsmittel wandelte, aber ebenso, dass die Airlines überhaupt zu den ersten Unternehmen gehörten, die sich von renommierten Designern und Graphikern ein umfassendes Corporate Design anpassen ließen. Und ebenso deutlich bestehen alle Gestaltungen aus einer Mischung von Markt- und Psychofaktoren, wobei die Balance zwischen beiden Funktionen das Entscheidende ist. In dieser Hinsicht war bei der Lufthansa die sicher einflussreichste und folgenträchtigste Designentscheidung, den Entwerfer Otl Aicher und die Entwicklungsgruppe 5 der Hochschule für Gestaltung Ulm 1962 das gesamte Erscheinungsbild der Lufthansa entwerfen zu lassen. Damit war das inzwischen in allen Branchen übliche Corporate Design erfunden. Dazu zählten Aicher und sein Team: „Hausfarben, Bildzeichen, Wortzeichen, Schriftart, Formate, grafische und typografische Ordnungen und Normen, Stil der Fotografie, Qualität der Trägermaterialien, Verpackungen, Ausstellungssysteme, Merkmale der Architektur, Formen (Gestalt) der Innenausstattungen und Geräte, Stil der Arbeits- und Dienstkleidung." Doch neben Aicher sind zahlreiche weitere renommierte Produkt- und Grafikdesigner, Modeentwerfer, Werbe- und Marketing-Agenturen für die Lufthansa tätig geworden. Zu ihnen zählen der Grafikdesigner Otto Firle, auf den das Kranich-Logo zurückgeht, die Produktdesigner Walter Dorwin Teague, Hartmut Esslinger und Frog, Priestman Goode, Jochen Müller und Jens Romca Industrial Design, Don Wollance, Wilhelm Wagenfeld, Hans Theo Baumann, Nick Roericht, Wolf Karnagel, Martin Topel & Herb B. Pauser, das Designbüro Bernd Hollin und Alexander Radoske, das Designbüro „Vision & Gestalt" (Bernd Bürdek, Stephan Schupach), die Modeentwerfer Uli Richter, Ursula Tautz, Werner Machnik, das Unternehmen Jobis, Jürgen Weiss und Gabriele Strehle mit ihrem Label „Strenesse" sowie die Agenturen Doyle/Dane/Bernbach, Zintzmeyer & Lux, die Peter Schmidt Group, Ogilvy & Mather, Young & Rubicam, Spiess/Ermisch/Abels, Fanghänel & Lohmann, Springer Jacoby und McCann-Erickson Communication House.

Natürlich haben auch die Gestaltungsabteilungen der Lufthansa selbst im Laufe der Jahrzehnte viele Produkte, Kampagnen und Gestaltungsstrategien für die Gesellschaft

passengers "on board" via the intercom system. The conductor is no longer called that but, picking up on the term "flight attendant," is now a "train attendant." The dining car has now become an "on-board restaurant" and train driver addresses passengers via the intercom as "your ICE captain." Many of the words are borrowed from the English terms used in flight operations: trolley, purser, counter, boarding, catering, the skytrain as an elevated railway serving the airport perimeter etc., to say nothing of the special Anglophile "dialect" used by pilots when talking with the tower and air traffic controllers.

All these bits of experience, which the airline uses when addressing passengers, draw upon a unique frame of reference for aviation, one which blends progress and nostalgia, internationalism and regionalism. Although an airline invariably bears the stamp of its home country, carriers always operate in an international arena and, for simple economic reasons, touch upon many cultures. Thus the Star Alliance was the first truly global alliance of airlines, founded in 1997 by the five founding airways – Air Canada, SAS, Thai Airways International, United Airlines and Lufthansa. Joining the alliance in the course of that same year were Varig, Ansett Australia, Air New Zealand and All Nippon Airways. Just like the Miles & More bonus program, the Star Alliance – now with more than 15 air carriers – is a "value added" system conceived at global scale.

When, in the pages which follow, we turn our attention to the design decisions taken in the past five decades at Lufthansa we will see that flying has been transformed from the exclusive pleasure of well-heeled passengers to an everyday means of mass transportation. We will also see that the airlines were among the very first companies to have a comprehensive corporate design concept worked out by renowned designers and graphic artists. And it is also quite clear that all the designs are based on a blend of marketing and psychological factors and that here the balance between these two functions is decisive. In this regard the most influential design decision at Lufthansa, the one with the greatest consequences, was made in 1962. It was then that Otl Aicher and "Development Group 5" at the Ulm College of Design were engaged to draft the entire corporate design identity for Lufthansa. Thus corporate design, which has in the meantime become common to all branches of industry, was invented. Aicher and his team included in their work the "corporate colors, symbols, logotypes, typeface, formats, graphic and typographical rules and standards, photographic style, quality of the printing papers, packaging, exhibit systems, features in architecture, shapes (design) of interior fittings and equipment, and style of professional attire." But in addition to Aicher, numerous other renowned product and graphic designers, fashion specialists, and advertising and marketing agencies have worked for Lufthansa. Among them was graphic designer Otto Firle, the creator of the crane logo. Others were product designer Walter Dorwin Teague, Hartmut Esslinger and Frog, Priestman Goode, Jochen Müller and Jens Romca Industrial Design, Don Wollance, Wilhelm Wagenfeld, Hans Theo Baumann, Hans "Nick" Roericht, Wolf Karnagel, Martin Topel and Herb B. Pauser, the Bernd Hollin and Alexander Radoske design office, the Vision & Gestalt design office (Bernd Bürdek, Stephan Schupach), fashion designers Uli Richter, Ursula Tautz, Werner Machnik, the Jobis company, Jürgen Weiss and Gabriele Strehle with her "Strenesse" label, along with various advertising and brand agencies – Doyle/Dane/Bernbach, Zintzmeyer & Lux, the Peter Schmidt Group, Ogilvy & Mather, Young & Rubicam, Spiess/Ermisch/Abels, Fanghänel & Lohmann, Springer Jacoby, and McCann-Erickson Communication House.

Walter Hanel: „Airport Athen", o. J.
Walter Hanel: "Athens Airport," undated

entwickelt. Aber die Auftragsvergabe an externe Designer eröffnet dann eben doch andere Möglichkeiten einer von außen entwickelten Produktsprache, die selbstverständlich dem generellen Erscheinungsbild der Fluglinie entsprechen sollte. Insofern sind präzise „Briefings“, Abstimmungsgespräche und die Heranführung der externen Designer an die logistischen Notwendigkeiten und Besonderheiten der Fluggesellschaft unabdingbar. Alessi oder Vitra können sich differente Autoren-Entwürfe leisten: ihre Unternehmensidentität besteht gerade aus der Diversifizierung unterschiedlichster Entwerfersprachen. Anders bei einer Fluglinie: Deren überwölbende „Corporate Identity“ muß alle Einzelprodukte integrativ betrachten, aufeinander beziehen und vernetzen. Dies heißt nicht, dass ein Erscheinungsbild und mit ihm die dazu gehörigen Produkte sich nicht weiterentwickeln könnten und sollten. Und so ist es ja auch: Bei der Lufthansa etwa hat es alle 5 bis 8 Jahre ein Relaunch gegeben, also bisher acht bis zehn Anpassungen an neue graphische und produktkulturelle jeweilige Gegenwarten. Auch muß man sich vor Augen führen, dass das Kerngeschäft einer Airline der Transport von Personen und Waren ist. So wichtig wie Farben und Logo, Geschäftspapiere und Produktdesign, Mode und Ausstattung auch sind: Sie können nicht das Herzstück ökonomischer Entscheidungen bilden. Aber sie bestimmen die Positionierung eines Unternehmens im Markt und gegenüber seinen Konkurrenten entscheidend mit: die Ästhetik des Bordgeschirrs, die Attraktivität der Dienstbekleidung oder der werbliche Auftritt sind nur einige dieser visuell wirksamen Elemente, welche die Publikumsansprache definieren und konturieren. Schon der Produktdesigner Raymond Loewy formulierte in den Mittdreißiger Jahren: „Hässlichkeit verkauft sich schlecht“ und dies ist bis heute so geblieben. Mehr als viele andere Airlines hat die Lufthansa dies berücksichtigt und von Anfang an ihr Corporate Design als strategisches Instrument im Wettbewerb verstanden. Es geht um größtmöglichen Komfort während des Fliegens, aber auch beim Ein- und Auschecken, beim Warten und Verweilen am Boden. „Fliegen heute“ ist zuallererst Dienstleistung und kaum noch mit dem Attraktivitätspotential des Abenteuers verbunden, welches vom Beginn des Fliegens in den 1910er Jahren bis in die 1960er Jahre anhalten sollte. Am Anfang eher etwas für unerschrockene Gemüter mit einem gewissen Hang zur Tollkühnheit, dann eher Privileg einer betuchten Elite, ist Fliegen heute – wie dies einmal ausgedrückt wurde – zum „komfortablen abgepolsterten Abenteuer“ geworden. Und: Fliegen ist auch ein Massengeschäft geworden, wie jeder „Urlaubsbomber“ nach Mallorca oder Antalya deutlich vor Augen führt. Gerade deshalb unternehmen in jüngerer Zeit Fluggesellschaften erneut beträchtliche Anstrengungen, um noch so etwas wie Distinktion herzustellen. Die Lufthansa hat jüngst in Frankfurt einen eigenen First-Class-Terminal eingeweiht, der an Exklusivität nichts zu wünschen übrig lässt. Dies alles sind imagefördernde Maßnahmen, aber nur ein kleiner Teil des ganzen Produkts. Und da wird in globalem Maßstab der Kampf um die Portionen härter. Die Zahl der Start- und Landeerlaubnisse (sog. „Slots“), Nachtflugverbot, politische Einflussnahmen, Kauf, Wartung, Logistik der Flugzeuge, Personal, Gewerkschaft, Arbeitnehmervertretungen, Abhängigkeit von schwankenden Ölpreisen, Lärmschutzüberlegungen, überhaupt Umweltfragen im weiteren Sinne sind nur einige Stichworte, die den Entscheidungshorizont einer Airline und eben auch ihren Problemhorizont andeuten. In diesem Gesamtkontext sind gestalterische Fragen ein Subkontext: Allerdings, wie ich meine, von nicht zu unterschätzender Bedeutung.

Im folgenden werden die gestaltungsrelevanten Produkte und Konzepte der Deutschen Lufthansa in „Gegenstandsfeldern“ aufgeblättert, um die entsprechenden Entwicklungen über die Jahrzehnte hinweg in den verschiedenen Gattungen zu verdeutlichen. Der Leser wird also immer wieder neu auf bestimmte Zeiträume stoßen. Er

Of course Lufthansa's own in-house design departments have developed many products, campaigns and design strategies over the course of the years. Outsourcing design tasks to external designers, however, opens the door to options for a product language developed from an outsider's viewpoint but always, of course, in alignment with the airline's overall identity. Thus precise briefings, coordination conferences and introducing outside designers to the logistical necessities and peculiarities of the airline are indispensable. Alessi or Vitra can afford to have differing designers' signatures show through; their corporate identity is, after all, based precisely on the diversification offered by a variety of vernaculars. The situation differs for an airline. Its overriding corporate identity has to observe and integrate all the individual products; they have to reference each other and form a network. This does not mean that the identity and the products associated with it cannot and should not be further developed. And that is exactly what has happened in the meantime. At Lufthansa there has been a relaunch every five to eight years and that means eight to ten modification cycles to accommodate new trends in graphics and product cultures. One must also keep in mind that an air carrier's core business is transporting passengers and cargo. As important as the colors and logo, business papers and product design, fashions and fittings might be – they can never be the touchstones for business decisions. They do, however, determine a company's position within the market and in relation to competitors. The esthetics of the dinnerware on board, the attractiveness of the uniforms and the presence of advertising in the public eye are just a few of these elements with visual impact which help lend definition and contours to the way in which the public is addressed. Product designer Raymond Loewy, in the mid-thirties, put it this way: "Never leave well enough alone." And this bit of wisdom has remained true down to the current day. More than other airlines, Lufthansa has taken this to heart and understood its corporate design to be a changing, strategic instrument in the competitive arena. The issues here are the greatest possible degree of passenger comfort – not only in flight but also when checking in and retrieving luggage and wherever a period of waiting is involved. Today flying is firmly footed in the services sector and can barely hold a candle to the appeal of adventure which prevailed from the start of aviation in the second decade of the last century and continued into the 1960s. Initially, manned flight was something more for imperturbable spirits with a certain penchant for daringness and subsequently more the privilege of a well-heeled elite. Today flying – as it was once said – has become a "comfortable and well-upholstered adventure." And flying has also become a high-volume business, as every "tourist bomber" that takes off for Mallorca or Antalya clearly demonstrates. That is one reason why carriers have recently undertaken renewed and considerable efforts to create something akin to distinction. Lufthansa recently opened a First Class terminal in Frankfurt and it leaves nothing to be desired in the way of privilege. These are all efforts intended to promote the airline's image, but are only a small part of the product as a whole. And at a global scale the battle for individual slices of the pie is becoming tougher. The number of slots available, banning night flights, political influence, purchasing, maintenance, airplane logistics, personnel, unions, employees' representatives, dependency on fluctuating oil prices, noise abatement needs and environmental issues in the broader sense are just a few of the key words which help to clarify the decision-making horizon for any airline and, at the same time, the range of problems which it has to address. In this overall context design questions are a subcontext but nonetheless one which, in my opinion, is of significance which should not be underestimated.

Walter Hanel: „Bombay", o. J.
Walter Hanel: "Bombay," undated

bleibt daher aufgefordert, die Querverbindungen zwischen den Kapiteln selbst herzustellen.

Schon 1919 hatte Otto Firle das Firmenzeichen des „Kranichs“ entworfen, welches trotz einiger Überarbeitungen bis heute das Unternehmen symbolisiert.

Unser Überblick über die 50-jährige Designgeschichte der Deutschen Lufthansa beginnt erst mit der Wiederaufnahme des Linienluftverkehrs 1955. Ab diesem Zeitpunkt lassen sich in historischer Abfolge im wesentlichen fünf Phasen der Designentwicklung bei der Deutschen Lufthansa feststellen.

Die erste Phase ab 1955 war zunächst durch eine sehr freie Werbegrafik mit heiterer, sympathischer Anmutung gekennzeichnet, auch wenn die blau/gelbe Farbigkeit durchaus schon zur Positionierung der Luftgesellschaft eingesetzt wurde. Die Wiederaufbau- und Wirtschaftswundereuphorie der späten 1950er Jahre zeigte sich aber auch im Selbstverständnis der neuen Lufthansa: heiter, demokratisch, unbeschwert, einen fröhlichen Optimismus ausstrahlend.

Diese von der Fremdwahrnehmung aus zurückstrahlenden Einschätzungen der Dominanzwerte der Deutschen Lufthansa wie technische Kompetenz, fliegerisches Können, Pünktlichkeit und Souveränität sollten dann auch in einem dazu kompatiblen Erscheinungsbild seinen Ausdruck finden. Und so professionalisiert sich ab 1962 mit der Beauftragung von Otl Aicher und seinem Team von der HfG Ulm in enger Zusammenarbeit mit der zeitgleich entstehenden Werbeabteilung der Lufthansa das gesamte Erscheinungsbild gravierend. Dazu zählten sämtliche visuelle Mittel des Unternehmens wie die Wort-/Bildmarke, überhaupt die Schriftzüge und die Schriftarten, der Satzspiegel, die Formate und Hausfarben, die Art der Fotografie, die Werbung, aber eben auch die Uniformen der Flugbegleiterinnen, die Bordgeschirre und die Ausstattung der Maschinen sowie der Stadtbüros, der Counter und Lounges am Boden. Die von Aicher und seinem Team entwickelte Corporate Identity galt lange Jahre auch international als führend und vorbildlich. Allerdings nahm Aichers Einfluß innerhalb der Lufthansa über die Jahre hinweg stark ab. Die von ihm bevorzugte funktionsorientierte Gestaltung bei extrem kontrollierter Emotionalität führte eben durchaus auch zu einem Eindruck von Kühle, Understatement und dominant bloß rationaler Kompetenz. Die meisten Weiterentwicklungen der späten 1960er und der 1970er Jahre wurden denn auch hausintern von der eigenen Werbeabteilung betrieben. Die Definition der einzelnen Designelemente und ihr systematischer Einsatz standen im Vordergrund. Mit der Wandlung von den regulierten Märkten im Flugverkehr zu liberalisierten Wettbewerbsverhältnissen wurden die definierten Designstandards allerdings sehr stark in Frage gestellt. Nun geriet die allzu dogmatische, unflexible, absolute „Prinzipienreiterei“, die vermeintlich nicht mehr den Anforderungen an markt- und zielgruppengerechte Kommunikation entsprach, in den Fokus der internen wie externen Kritik. Zum Teil gab es tatsächlich gravierende Fehlentwicklungen, etwa die schleichende Abschaffung der Bildmarke „Kranich im Kreis“, die strategisch gesprochen von sehr großem Nachteil war.

Walter Hanel: „München–St. Moritz“, o. J.
Walter Hanel: “Munich–St. Moritz,” undated

In einer dritten Phase wurde in den 1980er Jahren die schweizerische Corporate-Identity-Agentur Zintzmeyer & Lux mit einer zeitgemäßen Aktualisierung des Erscheinungsbildes beauftragt. Nachdem die Marktforschung immer klarer die Defizite des bisherigen Lufthansa-Images beschreiben konnte – so wurde etwa die Fluggesellschaft als reine Business Airline gesehen und fand kaum Akzeptanz in den Dienstleistungsdisziplinen – sollte mit Unterstützung der Zürcher Agentur ein breit angelegter Identitätsfindungsprozeß vollzogen werden. Daraus war gemäß dem Auftrag an die Agentur dann auch die Neuausrichtung des Erscheinungsbildes abzuleiten. Die Prämissen für die Gestaltung wurden nun weniger nur nach ästhetischen, sondern vielmehr nach wirkungs-

In this volume we will leaf through the many products and concepts at Lufthansa which are relevant to design. We will view them sector by sector to make clear the differing development vectors followed over the decades in the various sectors. The reader will encounter certain time frames again and again in the various chapters and I will leave it to the reader to make the cross references among chapters.

It was as early as 1919 that Otto Firle drafted the crane for use as the company emblem and, in spite of some revisions over the years, it continues to symbolize the company down to this very day.

Our review of the fifty years of design history at Deutsche Lufthansa begins with the resumption of scheduled service in 1955. Starting with that memorable date we can discern essentially five historically sequential phases in design development at Deutsche Lufthansa.

The first phase, commencing in 1955, was characterized initially by very free-spirited advertising graphics with a cheerful and appealing feeling while the combination of yellow and blue was used to position the airline in the market. The euphoria which typified the years of reconstruction and the *wirtschaftswunder* of the late 1950s was also reflected in the character of the "new" Lufthansa: cheery, democratic, unencumbered and radiating sunny optimism.

The assessments of the dominant values associated with Lufthansa, reflected back from outside observers, included technical competence, aviation expertise, punctuality and primacy in the field. These were to find expression in a corporate identity compatible with those values. The overall identity efforts were made significantly more professional as of 1962 when Otl Aicher and his team at the College of Design at Ulm were commissioned to collaborate with the Lufthansa Advertising Department, which came into being at the same time. Included here were all the visual means employed by the company, such as the logotype and emblem, lettering and typefaces, page layouts, formats and corporate colors, photographic style, advertising and even the flight attendants' uniforms, the airline dinnerware and the interior appointments, along with the sales offices and airport counters and lounges. The corporate identity developed by Aicher and his associates was for years internationally acclaimed as a leading model for such work. Aicher's influence within Lufthansa nonetheless dwindled over the course of the years. The function-oriented design which he preferred, accompanied by very closely controlled emotions, gave rise to a feeling of coolness and understatement and was dominated by simple, rational competence. Most of the further developments in the late 1960s and the 1970s were then carried through in-house by the firm's own advertising department. The definition of the individual design elements and their systematic employment were in the foreground.

With the transition from a tightly regulated airline industry to a more competitive operating setting, the design standards previously defined were earnestly questioned. The previous, rather dogmatic approach, far too inflexible and absolute, no longer seemed to represent a communications concept appropriate to addressing specific markets and target groups. Thus it became the subject of both internal and external criticism. There were in some cases grave aberrations such as the gradual elimination of the "crane in the circle" emblem, and this was of great disadvantage from the strategic viewpoint.

In a third phase, in the 1980s, the Swiss corporate identity agency Zintzmeyer & Lux was charged with updating the identity to conform with more contemporary standards. Market research ever more clearly revealed the deficits in the previous Lufthansa image. The carrier was seen as a pure business airline, for instance, and

bezogenen Kriterien überdacht. Denn aus der Ulmer Entwurfsauffassung des auch so genannten „Systemdesigns" entstand, ohne dass dies je beabsichtigt war, eine weitgehende Verwechselbarkeit der Erscheinungsbilder jener Unternehmen, die mit diesen Entwurfsrichtlinien (und fast immer von Aicher beraten) arbeiteten: Erco und Braun, BMW und FSB, das Institut für Auslandsbeziehungen … Zintzmeyer & Lux ging es zunächst darum, die Persönlichkeitsmerkmale des Unternehmens herauszuarbeiten. Sodann wurden davon Produkt- und Servicequalitäten wie die Zuverlässigkeit und Sicherheit, die technische Kompetenz und Pünktlichkeit, kundenfreundliche Gestaltung und Übersichtlichkeit der Flugpläne usw. abgeleitet. Erst danach erfolgte die Umsetzung dieser „Identität" in den Bereichen der Unternehmenskultur, dem sprachlichen Verhalten aller Mitarbeiter sowohl intern wie gegenüber den Medien und in moderaten Veränderungen des Erscheinungsbildes. Insgesamt war das Ziel dieses Ansatzes, die Unverwechselbarkeit des Unternehmens und der Marke zu stärken. Nun wurde das umfassende „Corporate Design" als wichtige Disziplin der Unternehmenskommunikation verstanden und als ein dominantes Medium zur Bewusstmachung von Qualität und Leistung nach innen und nach außen.

Als vierte Phase lässt sich eine Weiterentwicklung des Corporate Designs begreifen, die am Beginn der 1990er Jahre durch eine noch deutlicher serviceorientierte Ausrichtung des Unternehmens stattfand. Ende 1997 wurde die Deutsche Lufthansa schließlich vollständig privatisiert. Die damit einhergehende Intensivierung des freien Wettbewerbes führte zu einer größeren Emotionalisierung, die ihren Niederschlag in der Beauftragung des Designbüros Frog Design fand. Dieses Büro gestaltete neue Counter und Lounges, Flugzeugkabinen und Terminals. Das Reisen selbst als Erlebnis sollte in den Vordergrund gestellt und erfahrbar gemacht werden, um den Dienstleistungscharakter des Unternehmens noch stärker hervorzuheben. Insgesamt wurde das Produktdesign emotionaler mit weicheren zum Kunden hin geöffneten und zugewandten Formen und einem freieren und stärkeren Einsatz der Lufthansafarben als bisher. Im Rahmen eines von Frog Design so genannten „Retro-Futurismus" ging damit eine Besinnung auf die Ursprünge einher, auf den Pioniergeist und das Fliegen als Abenteuererlebnis der frühen Vorkriegszeit. Visualisiert wurde diese Rückbesinnung durch den dominanten Einsatz von Wellblech, welches das legendäre Junkers-Flugzeug gekennzeichnet hatte. Hartmut Esslinger, der Leiter dieses Designbüros, setzte die Zeichenhaftigkeit dieses Materials bewusst ein, um sowohl die Tradition wie die Zukunft des Unternehmens zu visualisieren. Eine solche retro-futuristische Produktsprache übertrug das Designbüro auch auf die Flugzeugsitze in allen Klassen.

Schließlich ist die fünfte, noch anhaltende Phase gekennzeichnet durch ein vorsichtiges Zurücknehmen dieser überbordenden emotionalen Produktsprache des kalifornischen Frog-Büros zugunsten einer sinnlichen High-Touch-Materialanmutung, die gewissermaßen eine gelassene Souveränität auszeichnet. Dabei stehen edle Naturmaterialien wie Hölzer und Leder gleichberechtigt neben High-Tech-Gestaltungen, etwa dem neuesten Flugzeugsitz des Londoner Designbüros Priestman Goode, einer veritablen High-Tech-Maschine, bei der die Lehnen in kühnem Parabelschwung bis zum Boden reichen. In der Business Class kann dieser Sitz zu einem luxuriösen Zweimeter-Bett werden. Insgesamt hat sich die Syntax der Kundenansprache weiter segmentspezifisch differenziert. Es geht sowohl um ein Designleitbild wie aber auch um die Pflege der „Marke" Lufthansa und das Herausstellen ihrer materiellen und ideellen Werte. Die gegenwärtig damit befasste Branding-Agentur Peter Schmidt Group sorgt sich in enger Abstimmung mit der Fachabteilung der Lufthansa also nicht nur um visuelle Zusam-

had poor scores in regard to service. A broad-based identity definition process was to be carried out with the support of the Zurich-based agency and, as specified in the commission, the results were to be used in the reorientation of the corporate image. The premises for design were to be shaped less in accordance with aesthetic criteria but rather with a view toward the impact of the design. This was due in part because the design philosophy adhered to in Ulm had unintentionally given rise to the so-called "systematic design concept" which in turn resulted in a certain sameness among the corporate images for the companies which worked with those design guidelines (and almost always using Aicher as a consultant). These included Erco and Braun, BMW and FSB, the Institute for Foreign Relations … The first task approached by Zintzmeyer & Lux was to work out the company's personality. It was from this that product and service qualities were derived, including reliability and safety, technical expertise and punctuality, customer-friendly design and clarity of the timetables, etc. Only then did the implementation of this "identity" follow in regard to corporate culture, the language used by the employees both internally and in relations with the media, and in moderate changes to the corporate image. All in all, the objective of this undertaking was to strengthen the unique position of the company and the brand. Comprehensive corporate design was now understood to be an important discipline in corporate communications and as a dominant medium for instilling awareness of quality and performance, both inside and outside the company.

Further refinement of the corporate design can be seen as the fourth phase. It commenced in the early 1990s and was marked by a clearer service orientation within the airline. At the end of 1997 the Deutsche Lufthansa was fully privatized. The intensification of market competition which this involved resulted in greater emphasis on emotions, and this culminated in turning to the Frog Design office. The agency created new counters and lounges, passenger cabins and terminals. The experience of travel was to be moved to the foreground and made easier to grasp. This was done in order to emphasize even further the service character of the company. Product design in general became more emotional and used softer forms, more open and directed to the customer. The Lufthansa colors were used more liberally than in the past. In the framework of the so-called "retro-futurism" embraced by Frog Design came the renascence of the company's roots, recalling the pioneering spirit and the adventure of aviation in the years just after World War I. This was visualized with the dominant use of corrugated sheet metal, which had characterized the legendary Junkers aircraft models. Hartmut Esslinger, the head of this design office, consciously used the symbolism of this material in order to visualize both the traditions and the future of the airline. His associates applied this retro-futurist product language to the aircraft seats for all classes of service.

Finally the fifth phase, still prevailing today, is characterized by a cautious reduction of the extravagantly emotional product vernacular of the Frog Office in California in favor of a sensual, high-touch character for the materials, embodying to a certain extent a sense of serene sovereignty. Here gracious natural materials including woods and leather take their place alongside high-tech concepts such as the newest airline seating, crafted by the Priestman Goode design office in London. This is a veritable high-tech seating machine in which the arms reach down to the floor in a daring parabolic curve. In the Business Class this seat can be converted into a luxurious two-meter bed. All in all the syntax of customer address has been further differentiated by segment. Here one is dealing not only with design guidelines but also the maintenance of the Lufthansa brand and the projection of its material and ideal values. The Peter

menhänge im Rahmen eines Corporate Design, sondern genereller auch um die Markenführung und um die sogenannte „Markenarchitektur".

Ziel einer solch breitgefächerten Darstellung ist es vor allem aber auch, die Entwicklung der Marke „Deutsche Lufthansa" zu beleuchten und ihre Markenwerte zu verdeutlichen.

Insgesamt müssen alle Anstrengungen im Zusammenhang der Markenführung und des Corporate Design darauf ausgerichtet sein, Glaubwürdigkeit und Authentizität, auch das persönliche Engagement aller Mitarbeiter für die Wünsche und Belange der Flugliniengäste zu fördern und zu steigern. Dafür hat sich die Deutsche Lufthansa auch im Sinne einer Selbstverpflichtung das langfristig gültige Motto gewählt: „There's no better way to fly."

Ich danke der Corporate-Design-Abteilung der Deutschen Lufthansa für die reichhaltige Bereitstellung von Bilddokumenten und Informationen. Selbstverständlich ergaben sich im Laufe der Recherche „Lücken", die es zu füllen galt, obwohl der Dokumentationsbestand allein über 3500 Bilder umfasst. Auch ein solches Weltunternehmen hat in seinen Dokumentationsanstrengungen unvermeidbare Reibungsverluste. Niemand kann mehr im Rückblick auf die Jahrzehnte alles wissen, aber man kann dann doch bei vielen, sich noch im Dienst befindlichen oder längst ausgeschiedenen Mitarbeitern Erinnerungen und Kenntnisse reaktivieren. Jedes Unternehmen hat ein kollektives Gedächtnis, welches aus den Erfahrungen und Erinnerungen der Mitarbeiter besteht. Auch diese Erinnerungen wurden genutzt. Sowohl Georg Schubert und Benita Struve, als auch Ralf-Dieter Rudorf und Ronald Wild von der Lufthansa-Abteilung „Markenmanagement" und „Corporate Design" waren mir ebenso unkonventionelle wie kompetente Partner und Ratgeber, die zudem ihrerseits sich nicht scheuten, den Rat der verschiedenen Fachabteilungen des Unternehmens einzuholen. Zusätzlich hat Christine Ramme in den verschiedenen Archiven der Lufthansa in Köln und Frankfurt am Main vieles zutage befördert, was in diese Darstellung eingeflossen ist. Auch die 2004 gezeigte Ausstelllung „Airworld – Design und Architektur für die Flugreise" des Vitra Design Museums in Weil am Rhein hat mir viele Anregungen gegeben wie auch das Buch von Keith Lovegrove: *Airline. Identity, Design and Culture*, New York / London 2000. Ebenso danke ich der Bibliothekarin des Rat für Formgebung in Frankfurt am Main, Frau Helge Aszmoneit, für manche wertvollen Hinweise. Mein Dank gilt ebenso den zahlreichen Redakteuren der Fachzeitschriften, aus deren Berichten ich vertiefende Informationen gewinnen konnte. Alexander Buck und Friederike Biegel von der Agentur Peter Schmidt Group vermittelten mir Einsichten in ihre Tätigkeiten für die Lufthansa, Sabine Runde vom Museum für Angewandte Kunst in Frankfurt am Main ließ mich an ihren Kenntnissen im Modesektor teilhaben. Aber ohne die aufmunternde, kritisch-solidarische Begleitung des Projektes durch meine Frau Anna Meseure-Fischer wäre dieses Buch mit Sicherheit ein anderes geworden, wenn es denn überhaupt entstanden wäre.

**1** Otl Aicher: „flugapparate von paul mac cready" (1981), Wiederabdruck in: *form*, Nr. 199, Nov./Dez. 2004, S. 53 ff.
**2** Hasso Spode: „Nichts wie weg hier! – Luftfahrt und Tourismus in historischer Perspektive", in: *Airworld – Design und Architektur für die Flugreise*, Vitra Design Museum, Weil am Rhein 2004.

Schmidt Group branding agency currently under contract works in close coordination with the departments at Lufthansa not only to tend to the visual contexts within the framework of the corporate design. In more general terms it deals with brand management and brand architecture.

The goal of such a broad-based depiction is, above all, to illuminate the development of the "Deutsche Lufthansa" brand and to make clear the value of the brand. In the broad view, all the efforts undertaken in conjunction with brand management and corporate design are aimed at promoting and increasing credibility and authenticity and encouraging the personal commitment of each and every employee to satisfy the wishes and needs of the airline's guests. To this end, and in the spirit of a voluntary pledge, Lufthansa has adopted as its long-term corporate motto: "There's no better way to fly."

I want to express my thanks to the Corporate-Design department at Deutsche Lufthansa for the rich availability of photographic documents and other information. Naturally, during the course of the research, gaps were discovered and they had to be filled, even though the document inventory contains more than 3,500 photos. There are also unavoidable "losses due to friction" in the documentation efforts for a global company such as this. No single person today can know everything in a review of the past decades. Memories and knowledge held by many employees still in service and those who have long since retired can nonetheless be reactivated. Every company has a collective memory, which is made up of the employees' experiences and memories. These memories were used, as well. Georg Schubert and Benita Struve, along with Ralf-Dieter Rudorf and Ronald Wild of Lufthansa's Brand Management and Corporate Design Department, were partners and advisors who were as unconventional as they were competent. They never hesitated to obtain advice from the various departments in the company. In additional, Christine Ramme brought a great deal to light at the Lufthansa archives in Cologne and Frankfurt and much of that is to be seen in this review. An exhibit staged in 2004 by the Vitra Design Museum in Weil am Rhein – "Airworld – Design und Architektur für die Flugreise" – was the source of many ideas. A further inspiration was a book by Keith Lovegrove: *Airline. Identity, Design and Culture*, New York/London, 2000. I also want to thank the librarian, Helge Aszmoneit, at the German Design Council in Frankfurt am Main for many valuable hints. My thanks also goes to the numerous editors of the professional journals; I was able to derive in-depth information from their articles. Alexander Buck and Friederike Biegel of the Peter Schmidt Group provided me with insights into their work for Lufthansa; Sabine Runde from the Museum for Applied Art in Frankfurt am Main permitted me to make use of its wealth of knowledge in the field of fashions. But without the encouraging, critical but nonetheless indulgent support of my wife, Anna Meseure-Fischer, this book would certainly have been quite different, if it had in fact come into being at all.

**1** Otl Aicher: "flugapparate von paul mac cready" (1981), reprinted in *form*, no. 199, Nov./Dec. 2004, pp. 53 ff.
**2** Hasso Spode: "Nichts wie weg hier! – Luftfahrt und Tourismus in historischer Perspektive," in: *Airworld – Design und Architektur für die Flugreise*, Vitra Design Museum, Weil am Rhein, 2004.

# „man ist so, wie man sich zeigt, und wie man sich zeigt, so ist man“ (Otl Aicher)

Entgegen einem weit verbreiteten Missverständnis ist „Corporate Design“ nicht das gleiche wie „Corporate Culture“. Während der erste Begriff alle visuell relevanten Maßnahmen meint, also das graphische und produktkulturelle Umfeld nach innen und außen, beschreibt der zweite auch das Arbeitsklima der Beschäftigten untereinander. „Corporate Culture“ ist damit auch ein Begriff, der soziologische und psychologische Zusammenhänge anspricht und im Kern die intellektuelle Ausrichtung eines Unternehmens nach innen und außen umfasst. Denn zur Kultur eines Unternehmens gehört nicht nur sein Umgang mit den Interessen, Bedürfnissen und Belangen aller Mitarbeiter, also mit dem, was der Volksmund „Betriebsklima“ nennt, sondern auch ihr Umgang mit Kunden, Vertriebspartnern, Lieferanten und generell der Öffentlichkeit. Zudem gibt es die Begriffe „Corporate Identity“, „Corporate Behavior“ und „Corporate Communication“. Auch sie beschreiben Wahrnehmungsfelder innerhalb einer Unternehmensidentität. All diese Begriffe benennen Strukturen und Prozesse, die hohe Sensibilität erfordern. Sowohl die Entwicklung eines Erscheinungsbildes wie dann im Laufe der Jahre und Jahrzehnte seine Veränderungen sind zuallererst Elemente des Selbstverständnisses eines Unternehmens, treffen deshalb bei den Mitarbeitern immer auf reges Interesse und lösen manchmal durchaus erregte Debatten aus.

Das graphische Erscheinungsbild der Lufthansa begann wie bei vielen anderen Unternehmen auch mit der Entwicklung einer Bild- und Wortmarke. Otto Firle, von Hause aus Architekt, entwarf 1918, ein Jahr vor Beginn des regelmäßigen weltweiten Passagierflugverkehrs, für die Deutsche Luft Reederei das Kranich-Zeichen, das, wie die Zeitschrift *Architektur & Wohnen* einmal schrieb, heute „der verbreitetste Großvogel am Himmel über Deutschland“ sein dürfte. Anfang 1926 fusionierte die Deutsche Luft Reederei mit der Junkers Luftverkehr AG zur Deutschen Luft Hansa Aktiengesellschaft. Von Junkers wurden die Farben Blau und Gelb für die neue AG übernommen. Das Signet des langgestreckten, aufsteigenden Kranichs erschien am Anfang in dreifacher Kombination und symbolisierte damit nicht zuletzt auch einen strengen Formationsflug. Aber bereits ein paar Jahre später wird dieser fast weltweit verbreitete weiße Großvogel, den eine ähnliche, hochbeinig-langhalsige Eleganz wie einen Flamingo auszeichnet, einzeln im Kreis verwendet. Auch nach dem Neuanfang der Lufthansa 1955 bestimmten der Kranich und der zurückhaltende Einsatz der Farben Blau und Gelb das äußere Erscheinungsbild der Fluggesellschaft. Keith Lovegrove bilanziert:

„Fast alle Fluggesellschaften haben eine Bildmarke; eine Gewohnheit, die bereits 1919 begann, als der Vorgänger der Lufthansa, die Deutsche Luft Reederei, das Motiv eines fliegenden Kranichs als Logo installierte, welches bis heute, wenn auch behutsam modifiziert, den Kern der historischen Identität der Lufthansa bildet. Ein Premium-Image bei Geschäfts- und Privatreisenden zu erreichen, ist das Ziel jeder Fluggesellschaft und ein bloß gelungener grafischer Auftritt alleine reicht da nicht. Die Bildmarke alleine ist nur ein kleiner Teil des gesamten Prozesses der Entwicklung einer Corporate Identity. Ebenso wichtig ist es, die Identität und das Image der Fluggesellschaft über die Bordausstattung, die Flughafenlounges und die Gates, über ihr Schrifttum, die Flugscheine, die Uniformen und Bodenfahrzeuge zu vermitteln. Zusätzlich zu allen gestalterischen Überlegungen muss gleichberechtigt eine nach innen vermittelbare Management-Haltung und ein deutlich freundlicher Kundenservice treten.“ Und bemerkte darüber

Otto Firle: Plakat für die Deutsche Luft Reederei, 1919
Otto Firle: Poster for the Deutsche Luft Reederei, 1919

# "You are what you show, and what you show reveals what you are." (Otl Aicher)

Contrary to widespread misunderstanding, "corporate design" is not the same as "corporate culture." Whereas the first concept covers all the relevant visual aspects – including the graphic environment and product culture both internally and externally, the second describes, in addition, the working atmosphere among the employees. "Corporate culture" is thus also a concept which deals with sociological and psychological contexts. In its essence, it comprises the overall intellectual thrust of a corporation, both inwardly and outwardly. This is because a corporation's culture embraces not only its dealings with the interests and needs of all its employees – sometimes referred to as the work atmosphere, but also its dealings with customers, suppliers, government authorities and the general public. The terms "corporate identity," "corporate behavior" and "corporate communication" are also used. They, too, describe fields of perception within a corporation's overall identity. All these terms designate structures and processes that require a high degree of sensitivity. Both the initial development of the corporate image and its changes in the course of subsequent years and decades are, above all, expressions of a company's own view of itself. This is why changes are always met with lively interest among employees and sometimes provoke heated debates.

As with so many other firms, Lufthansa's graphic identity first emerged with the development of a company emblem and logotype. In 1918 Otto Firle, architect by trade, drafted the crane symbol for the Deutsche Luft Reederei, one year prior to the commencement of scheduled, worldwide passenger air traffic. The magazine *Architektur & Wohnen* once wrote that the crane might well today be "the large bird most commonly seen in the German skies." At the beginning of 1926 the Deutsche Luft Reederei merged with Junkers Luftverkehr AG to form the Deutsche Luft Hansa Aktiengesellschaft. This new, publicly traded company adopted Junkers' corporate colors, blue and yellow. The stretched, soaring crane initially appeared in a group of three, suggesting strict formation flight. It was not but a few years later that this large white bird, found almost everywhere on the planet and distinguished by a long-legged, long-necked elegance similar to that of a flamingo, was used singly, within a circle. After the revival of the Lufthansa organization in 1955 the crane and the restrained use of the colors blue and yellow once again placed their stamp on the airline's public image. Keith Lovegrove summarized:

"Most airlines have a logo, a convention that began in 1919 when a predecessor of Lufthansa, the Deutsche Luft Reederei, applied a flying crane motif that in only a slightly modified form is still the anchor for a very successful history of corporate identity at Lufthansa. Achieving premium-brand status, and thereby winning loyal custom from business and occasional travelers alike, is the goal for any airline and graphic design alone is not the solution. The logo and aircraft livery represent a small part of the identity process. Equal importance is attached to communicating the brand identity via aircraft interiors, airport lounges and gates, corporate literature, tickets, uniforms and ground vehicles. In addition to the aesthetic aspect, it is increasingly important to consider management ethos and customer service as part of the image review …

"Historically, few logos have survived the critical eyes of the marketing and design departments. Lufthansa, however, … has retained its flying crane since 1926." [1]

It continues to be important, in addition, that the solitary crane does not trigger military and heraldic connotations as would an eagle, for instance, but it is nonetheless

Seite 24: Otl Aicher: Flugzeugbemalung, Boeing 737-200, ab Ende 1960er Jahre
Page 24: Otl Aicher: Livery, Boeing 737-200, as of late 1960s

Erster Flugplan der neuen Lufthansa, 1955
First timetable for the new Lufthansa, 1955

hinaus: „Unter historischen Gesichtspunkten haben nur wenige Bildmarken über längere Zeiten hinweg den kritischen Einwänden der Marketing- und Designabteilungen standhalten können. Die Lufthansa aber hat ihren fliegenden Kranich seit 1926 beibehalten.“ [1]

Zusätzlich bleibt wichtig, dass der einzelne Kranich nicht heraldisch-militärische Konnotationen auslöst wie etwa ein Adler. Aber er ist majestätisch genug, um die Grandezza einer Fluggesellschaft zu symbolisieren. Die frühe Form des Kranichs zeigte jugendstilhaft-beschwingte Kurvaturen, die auch dem Nachkriegsjahrzehnt wieder aktuell erschienen. Auch die Wortmarke „Lufthansa“ aus Großbuchstaben wurde nun kursiv. Beides korrespondierte mit dem Formgefühl der Zeit, die Schwünge und organische Formen bevorzugte, weil jede klassizistisch wirkende Symmetrie an konservative Formensprachen erinnerte. Zudem war die amerikanische „Streamline“ eines Raymond Loewy, Henry Dreyfuss oder Harley Earl, die alle für amerikanische Fluglinien Produkte, graphische Erscheinungsbilder und Interieurs entwarfen, nicht nur formal ähnlich, z. B. bei der Lockheed Super Constellation, sondern auch als Formensprache des neu entdeckten Westens positiv besetzt. So wurde etwa die Super Constellation liebevoll „Super Conny“ genannt. Jochen Eisenbrand formulierte 2004 in der Ausstellung „Airworld“: „Aufgrund ihres elegant geschwungenen Rumpfes und ihres dreifachen Heck-Leitwerks gilt die Lookheed Super Constellation bis heute als eines der schönsten Passagierflugzeuge, das je gebaut wurde.“ [2]

Ansonsten aber war das Erscheinungsbild der Lufthansa in den Nachkriegsjahren visuell eher bunt und unsystematisch, wenn auch blau und gelb dominiert, allerdings mit der durchgehaltenen Parabellinie auf dem Leitwerk der Flugzeuge, den Gepäckschildern und Geschäftspapieren, den Flugplänen, Jahresberichten und den Bodenfahrzeugen, zu denen auch Traktoren gehörten, und selbst den Zigarrenbanderolen. Die oft zufällig wirkende Typografie und die illustrative Grafik, die bis zu Karikaturen reichte und etwa auf Notizblöcken betonte: „Fliegen ist billiger als man denkt!“ waren noch weit davon entfernt, die Systematik eines einheitlichen Auftrittes anzustreben.

Dann aber wurde alles anders. Ein ehemaliges Vorstandsmitglied schreibt:

„In der Geschichte des Erscheinungsbildes der Lufthansa ist das Jahr 1962 das Schlüsseljahr. In diesem Jahr verabschiedete der Vorstand das von Otl Aicher und seinem Team in Zusammenarbeit mit der Werbeabteilung entwickelte Erscheinungsbild der Lufthansa. Nach der Umsetzung war Lufthansa viele Jahre die ‚Design-Leading‘-Airline und gefragtes Muster-Unternehmen für zum Ausdruck gebrachte Corporate Identity – lange bevor dieser Begriff in Deutschland zum Schlagwort wurde.“ [3]

Aicher und sein Team firmierten unter dem Kürzel „E5“. Diese Arbeitsgruppe bestand aus Tomàs Gonda, Fritz Querengässer, Hans ‚Nick‘ Roericht, Alfred Kern und ihrem Leiter Otl Aicher. Die Gruppe arbeitete an der berühmten Hochschule für Gestaltung in Ulm, zu deren Gründungsmitgliedern Aicher zählte. Bernd Franck, der als HfG-Student damals an diesem Projekt mitarbeitete, erinnert sich:

„1962 hatte die Lufthansa die Gruppe mit der ‚Ausarbeitung von Richtlinien und Normen für visuelle Gestaltungsaufgaben der Deutschen Lufthansa‘ beauftragt. Darunter war zu verstehen: Feststellung des Ist-Zustands und Ausarbeitung eines umfassenden Soll-Zustands, mithin eines neuen Erscheinungsbildes. Das Projekt 1400 – wie die Lufthansa-Arbeit fortan hieß – wurde zunächst in einer Studie ausführlich beschrieben, sie war Basis für die weitere Vorgehensweise. Neben der eingehenden Auflistung und Präsentation des bisherigen Unternehmensauftritts hielten die Ulmer kapitelweise neue Gestaltungsansätze für ein umfassendes und einheitliches Erscheinungsbild fest.“ [4] Dazu zählten Aicher und sein Team:

majestic enough to symbolize the preeminence of the airline. The crane's early embodiment showed flowing curvatures in the art nouveau tradition and they once again seemed timely in the post-war decade. The Lufthansa logotype, all in capital letters, was now set in italics. Both corresponded to the esthetics of the era, giving preference to flowing strokes and organic forms, since any symmetry with a classical feel reminded viewers of more conservative design vernaculars. In addition, the American "streamline" concept embraced by Raymond Loewy, Henry Dreyfuss and Harley Earl, all of whom drafted products, graphic designs and interiors for U.S. airlines, was similar not only in its formal aspects – in the Lockheed Super Constellation, for instance. It also enjoyed positive associations with the design vernacular used in the newly discovered West. Thus, for instance, the Super Constellation was fondly referred to as the "Super Connie." At the "Airworld" exhibition staged in 2004 Jochen Eisenbrand summed up: "Owing to its elegantly curved fuselage and its triple rudder the Lockheed Constellation is deemed, down to this very day, to be one of the most attractive passenger aircraft ever built." [2]

Otherwise Lufthansa's visual image in the post-war years was more or less colorful and without a consistent system, even though blue and yellow were dominant. One constant was the parabolic line appearing on aircraft rudders, the baggage tags and corporate stationery, timetables, annual reports and ground vehicles – right down to the tow trucks, and indeed including cigar bands for complimentary stogies. The typography often seemed haphazard and the illustrative graphics, which even included cartoons, declared on notepads, for example, that "Flying is more affordable than you'd think!" All this was a far cry from the system employed to achieve a uniform image.

Then everything changed. A former member of the Board writes:

"1962 was the turning point in the history of the Lufthansa image. It was in that year that the Executive Board adopted the Lufthansa identity, developed by Otl Aicher and his team in cooperation with the advertising department. Following implementation, Lufthansa was for many years the design leader among airlines and a much-emulated trailblazer in the expression of corporate identity – long before this concept had become a household word in Germany." [3]

Aicher and his team operated under the abbreviation "E5." This working group was made up of Tomàs Gonda, Fritz Querengässer, Hans "Nick" Roericht, Alfred Kern and their chief, Otl Aicher. This group worked at the famed Design College in Ulm; Aicher was one of its founding faculty members. Bernd Franck, who was involved in the project as a student at that time, recalls:

"In 1962 Lufthansa commissioned the group with a project entitled 'Working out guidelines and standards for visual design tasks at the Deutsche Lufthansa.' The commission included determining the current status and working out a comprehensive target situation and then a new visual image. 'Project 1400' – as the Lufthansa commission came to be known – was initially described in detail in a study which served as the basis for further work. In addition to a detailed listing and presentation of the corporate image to date, the experts at Ulm codified, chapter by chapter, new design approaches for a comprehensive and uniform visual image." [4]

Included in this listing compiled by Aicher and his team were "corporate colors, symbols, logotypes, typeface, formats, graphic and typographical rules and standards, photographic style, quality of the printing papers, packaging, exhibit systems, features in architecture, shapes (design) of interior fittings and equipment, and style of professional attire." Thus detailed planning for the company's outward image had been invented. This has in the meantime come to be known in every branch of industry as "corporate identity."

Lockheed Super Constellation Flugzeuge, Flughafen Frankfurt, Ende der 1950er Jahre
Lockheed Super Constellation aircraft, Frankfurt Airport, late 1950s

Otl Aicher, Tomàs Gonda, Fritz Querengässer, Hans Roericht: Erscheinungsbild der Lufthansa, 1962, Auszug aus dem Manual
Otl Aicher, Tomàs Gonda, Fritz Querengässer, Hans Roericht: Lufthansa Corporate Identity, 1962, extract from the design manual

„Hausfarben, Bildzeichen, Wortzeichen, Schriftart, Formate, graphische und typografische Ordnungen und Normen, Stil der Fotografie, Qualität der Trägermaterialien, Verpackungen, Ausstellungssysteme, Merkmale der Architektur, Formen (Gestalt) der Innenausstattungen und Geräte, Stil der Arbeits- und Dienstkleidung."

Damit war die detaillierte Planung des Firmenauftrittes erfunden, die inzwischen in allen Branchen übliche „Corporate Identity".

Zunächst veränderten Aicher und sein Team die Farben: Das Gelb wurde wärmer, sonniger, das Blau dunkler, eher an das Meer als an die Wolken gemahnend. Sonne, Sand, Wellen, Meer. Die Farbänderungen nehmen subtil auch auf die steigende Zahl touristischer Urlaubsgäste Bezug:

„Die Hausfarben der Lufthansa – Blau und Gelb – wurden beibehalten, jedoch subtil modifiziert (das Gelb in Richtung Orange, das Blau ins Violett). Das Gelb sollte insgesamt in den Vordergrund treten. Das Konzept wurde verworfen, war dennoch grundlegend für die weitere Entwicklung."[5]

Dies heißt dann schlussendlich doch, dass das Konzept weitgehend vor allem bei den Drucksachen umgesetzt wurde. Bei der Bemalung der Maschinen schlug das Aicher-Team zunächst eine Wortmarke in gemischter Schreibweise vor, gesetzt aus der „Stempel Helvetica" im Schnitt „Helvetica halbfett". Um die Erscheinungsbildänderung nicht so hart ausfallen zu lassen, wurden zunächst einige Maschinen als Übergangsversion in Versalschrift bemalt, denn über lange Jahre war die Schrift „Schadow Antiqua" in Versalien zur angewöhnten Wortmarke der Lufthansa geworden. Das Leitwerk sollte vollflächig gelb werden, mit dem Signet des blauen Kranichs im blauem Kreis in seiner Mitte. Aber wie einem Alternativentwurf von Aicher aus dem Jahre 1963 für die Maschinenbemalung zu entnehmen ist, blieb das weiße Leitwerk mit der seit sieben, acht Jahren eingeführten Parabel auf Drängen des Lufthansa-Vorstands bestehen. Dieses modifizierte Erscheinungsbild wurde bis 1967 verwendet. In ihrer Projektstudie hatte die „Entwicklungsgruppe 5" akribisch den Fahrzeugpark, bestehend aus Flugzeugen, Bodenfahrzeugen wie PKWs, Bussen und Kleinlastwagen, Rollfeldtreppen sowie Zug- und Gepäckfahrzeugen, zunächst in deren Erscheinungsbild der Zeit von 1955 bis 1962 dargestellt, um diesem dann ihren eigenen Vorschlag gegenüberzustellen. Die Bodenfahrzeuge zeigten nun als Basisfarbe das Gelb, der Kranich im Kreis stand blau auf dem Gelb, die Wortmarke dagegen gelb auf blauem Band. Vor allem bei den typografischen Anwendungen kam die nüchterne, rationale Eleganz der Ulmer Hochschule zum Tragen. Otl Aicher erinnerte sich Jahrzehnte später:

„man ist so, wie man sich zeigt, und wie man sich zeigt, so ist man. das erscheinungsbild ist nicht nur das äußere, wenn es auch von vielen so nach der klassischen philosophie gesehen wird. es ist das eigentliche. man kann nicht existieren, ohne sich zu zeigen, und wie man sich zeigt, so ist man. bei der lufthansa war der gestaltungsvorgang nicht ein solcher der chefetage. dazu war das unternehmen zu groß und dem herkömmlichen system der arbeitsteilung unterworfen. es war nicht die gemeinsame intellektuelle und spekulative arbeit wie z. b. bei braun. der vorstand bestellte, entschied und ließ ausführen. das zeigte sich auch am resultat. dieses wurde zwar gut und besteht in der substanz bis heute, ein optimum war es nicht, wenngleich eine hervorragende designabteilung etabliert wurde …"[6]

Bereits 1961 wurde vom damaligen Lufthansa-Vorstand Hans G. Conrad (der ab 1953 einer der ersten HfG-Ulm-Studenten war) als Werbeleiter berufen. Vorher war er Leiter für Grafik Design bei der Firma Braun gewesen. Er war 1962 entscheidend für die LH-Beauftragung von Otl Aicher und seinem Team verantwortlich. Auch Klaus Wille als erster festangestellter Designer der Lufthansa und Hermann Roth kamen von der Firma

„man ist so, wie man sich zeigt, und wie man sich zeigt, so ist man. das erscheinungsbild ist nicht nur das äußere, wenn es auch von vielen so nach der klassischen philosophie gesehen wird. es ist das eigentliche. man kann nicht existieren, ohne sich zu zeigen, und wie man sich zeigt, so ist man. bei der lufthansa war der gestaltungsvorgang nicht ein solcher der chefetage. dazu war das unternehmen zu groß und dem herkömmlichen system der arbeitsteilung unterworfen. es war nicht die gemeinsame intellektuelle und spekulative arbeit wie z. b. bei braun. der vorstand bestellte, entschied und ließ ausführen. das zeigte sich auch am resultat. dieses wurde zwar gut und besteht in der substanz bis heute, ein optimum war es nicht, wenngleich eine hervorragende designabteilung etabliert wurde …“

“You are what you show, and what you show reveals what you are. The image is not just the outward appearance, even though many see it as such in accordance with classical philosophy. It is the essence. One cannot exist without showing oneself and the way one shows oneself, that is what one is. At Lufthansa the design proposal was not a product of the boardroom. The company was too large for that and it was subject to the conventional system of the division of labor. It was not joint intellectual and speculative work such as was pursued at Braun. The Executive Board awarded a commission, made a decision and had it carried out. This is also evidenced in the result. It was indeed good and, in substance, survives to the present day, but it was not the optimum, even though an excellent design department was established.”

**Otl Aicher, 1962**

Braun und wurden von Hans G. Conrad engagiert. Seit 1970 bis ca. 1985 war Ernst-Theodor Tschoepke Leiter der Werbeabteilung, der in dieser Zeit gemeinsam mit Klaus Wille die konsequente Fortführung der Design-Aktivitäten der Lufthansa ermöglichte. Auch Frank Cesarz war über lange Zeit als Direktor für Öffentlichkeitsarbeit an der gestalterischen Profilierung der Lufthansa entscheidend beteiligt.

In einer zweiten Stufe überarbeitete ab 1966/67 die interne Designabteilung der Lufthansa mit Mitarbeitern, die – wie gesagt – aus der Aicher-„Schule" stammten, das Erscheinungsbild. Einiges war so dicht am Denken des „Meisters", dass manche publizistische Berichte fälschlicherweise von Aicher als Autor ausgingen. Die interne Arbeitsgruppe konzipierte nun die Unterseite der Flugzeuge blank poliert und das Leitwerk blau mit blauer Bildmarke auf gelber Kreisfläche. Die Gruppe „LM3" hatte für diese Überarbeitung allein über 25 Leitwerkvarianten entwickelt, die keineswegs die Zustimmung von Aicher fanden. Bei den Bodenfahrzeugen rutscht der blaue Streifen mit gelber Wortmarke ein gutes Stück nach unten und lässt die Fahrzeuge dadurch eleganter erscheinen. Die von Aicher bevorzugte typografische Rationalität kam auch allen Tochtergesellschaften zugute und blieb selbst dann erhalten, wenn die Farbpaletten verändert wurden. Das Erscheinungsbild orientierte sich weiterhin an Aichers Systemdesign, auch wenn in den Farben, der Typografie und den Stilen der Fotografie zunehmend größere Abweichungen auftreten, etwa bei Stadtführern oder Länderplakaten. Zudem wurde ab 1970 das von Aicher entwickelte Passagierleitsystem eingesetzt. Jochen Eisenbrand erläutert:

„Mit dem Anstieg des internationalen Flugverkehrs wuchs in den 1960er Jahren der Bedarf an international verständlichen Passagierleitsystemen. Zu den ersten Fluglinien, für die eigene umfassende Zeichensysteme als Wegweiser entwickelt wurden, gehörten der Amsterdamer Flughafen Schiphol (Total Design, 1965) und der Flughafen in Frankfurt am Main (Otl Aicher, 1970)." [7]

Der „Übervater" Otl Aicher blieb dem Unternehmen selbst dann noch erhalten, als der Vorstand 1979 seinen neuen Vorschlag, die „Analyse 80", zur Auffrischung des Erscheinungsbildes ablehnte. Nun wurde die „Helvetica" in einem nochmals fetteren Schnitt verwendet, der „Helvetica demibold". Auch hier wieder bleibt die Typografie im Sinne eines durchgehenden Corporate Designs bei allen Tochtergesellschaften und Beteiligungen ebenso gleich wie bei allen Lufthansa-Geschäftspapieren, von Flug- und Angebotsdokumenten über Flugpläne und geprägte Service-Cards im Scheckkartenformat bis zu den Unterhaltungsprogrammen im Flugzeug für die über Kopfhörer zu empfangende Musik.

Boeing 707-430 Flugzeug mit „Schadow Antiqua"-Schriftzug und „Parabel"-Leitwerk, 1957–1963
Boeing 707-430 with "Schadow Antiqua" typeface and the "parabola" on the tail unit, 1957–1963

In the first step Aicher and his team modified the colors. The yellow became warmer and sunnier, the blue darker, reminiscent more of the ocean than of clouds. Sun, sand, waves, the sea. The color change made subtle reference to the rising numbers of vacationing passengers.

"The Lufthansa corporate colors – blue and yellow – were retained but subtly modified (yellow toward the orange and blue into the violet range). Yellow was in general to move into the foreground. The concept was rejected but was nonetheless fundamental to further development." [5]

This means that ultimately the concept was largely implemented, above all in printed matter. As regards painting the aircraft, the Aicher team first suggested a logotype with a single capital letter, to be set using the "Stempel Helvetica" typeface in the semi-bold family. In order to keep the change in the image from being overly abrupt, a few aircraft were initially painted in all-caps as a transitional version because, over a period of many years, the "Schadow Antiqua" typeface in all-caps had become the familiar logotype for Lufthansa. The rudder was to be entirely in yellow, showing at its center the blue crane emblem within a circle. But, as can be seen from an alternate proposal by Aicher for aircraft livery, dating back to 1963, the rudder remained in white and – at the insistence of the Lufthansa Executive Board – featured the parabola which had been introduced seven or eight years before.

This modified image was used through 1967. In its project study the "Development Group 5" described the airline's equipment in painstaking detail. The list included the aircraft themselves, ground vehicles such as passenger cars, busses and small vans, air stairs and tugs, and luggage-moving vehicles. This study initially documented the embodiments in the period from 1955 to 1962, in order to then contrast them with the Group's own proposal. The ground vehicles now used yellow as the basic color, against which was seen the crane in the circle, both in blue; the logotype differed, being in yellow on a blue stripe. Above all, the typographical applications benefited from the sober and rational elegance of the Design College at Ulm. Decades later Otl Aicher recalled:

"You are what you show, and what you show reveals what you are. The image is not just the outward appearance, even though many see it as such in accordance with classical philosophy. It is the essence. One cannot exist without showing oneself and the way one shows oneself, that is what one is. At Lufthansa the design proposal was not a product of the boardroom. The company was too large for that and it was subject to the conventional system of the division of labor. It was not joint intellectual and speculative work such as was pursued at Braun. The Executive Board awarded a commission,

Otl Aicher: Flugpläne, 1966
Otl Aicher: Timetables, 1966

Designabteilung der Lufthansa: Boeing 737 mit „Helvetica"-Schriftzug und Bildmarke auf gelbem Kreis, ab Ende 1967
Lufthansa design department: Boeing 737 with "Helvetica" logotype and emblem against a yellow circle, as of late 1967

Bereits 1978 hatte die Designabteilung der Lufthansa eine „Modifikation von LH-Schriftzug, -Firmenzeichen und -Hausfarben Gelb/Blau für die Anwendung bei Fahrzeug- und Flugzeugbemalungen sowie Beschriftungen von Stadtbüros, Stationen, Ausstellungen" durchgeführt und zwei neue RAL-Farben beantragt. Mit deutscher Gründlichkeit wurde diesem Antrag stattgegeben:

„Auf Antrag der Deutschen Lufthansa AG, Köln, sind zwei neue Farben in das Farbregister RAL 840 HR aufgenommen worden. Diese Farben sind ein wesentliches Element des bekannten Lufthansa-Emblems, das international einheitlich angewendet wird. Dazu sind entsprechende Farbvorlagen erforderlich, um auch bei Lieferungen verschiedener Hersteller einen möglichst gleichen Farbausfall sicherzustellen. Die Verwendung der Farben erstreckt sich auf alle Bereiche der visuellen Kommunikation, wie Flugzeug- und Fahrzeuglackierung, Außenanstriche, Plastikverarbeitung, Textileinfärbungen und Textildruck, Drucksachen usw. Zu den genannten Verwendungsbereichen der Lufthansa kommen außerdem noch die Tochter- und Beteiligungsgesellschaften, wie Condor, Hansa Luftbild, Lufthansa Service GmbH. Es wurde berücksichtigt, dass im Farbregister bisher keine vergleichbaren Farben enthalten sind und der Abstand zu ähnlichen im Register befindlichen Farben erheblich ist. Die neuen Farben haben folgende Farbbezeichnungen und Farbhilfsbezeichnungen erhalten: RAL 1028 Melonengelb, RAL 5022 Nachtblau."[8]

In der Fachzeitschrift *Design Report* hieß es 1989 im Rückblick auf Otl Aichers „Analyse 80":

„Vor einem Jahrzehnt begann die Lufthansa-Spitze über ihr visuelles Erscheinungsbild nachzudenken. Man wollte eine Revision des alten Outfits aus Ulm, jenes Erscheinungsbildes, das mittlerweile Designgeschichte ist. Otl Aicher hatte es zu Beginn der sechziger Jahre mit Hans ‚Nick' Roericht an der Ulmer Hochschule für Gestaltung konzipiert: blaues Farbband, gelbes Leitwerk mit dem blauen Kranich. Der Rumpf unten silberfarben und oben weiß. Der Lufthansa war es nach über zwanzig Jahren zu kühl, zu streng und vornehm. Nicht mehr ‚High-Tech', sondern ‚High-Touch' ist gefragt. Man will den neuen Kunden ‚emotional' ansprechen. So wurde Otl Aicher – Anfang der achtziger Jahre – mit der Überarbeitung seiner eigenen Entwürfe aus den sechziger Jahren betraut. Doch Aicher wollte allenfalls eine Revision in eigenen Maßen. Das erschien den Airlinern gar zu behutsam. Darum beauftragte die Lufthansa 1986 die Zürcher Corporate-Identity-Agentur Zintzmeyer & Lux, Konzepte für eine neue Gesamterscheinung zu finden."[9]

Offenbar jedoch sind die tatsächlichen Geschehnisse nicht immer identisch mit einer öffentlichen Berichterstattung. Denn bereits Mitte der 1980er Jahre hatte der Lufthansa-Vorstand beim amerikanischen CRC, dem „Communication Research Center", eine Studie mit dem Titel „Image 2000" in Auftrag gegeben. Bezogen auf die Corporate Identity der Airline, bilanzierte die Studie ein Defizit an Emotionalität sowie eine LH-Positionierung als reine Geschäftsreisen-Fluggesellschaft, mit der Hochpreisigkeit und elitäres Image verbunden werde. Demgegenüber werde der Faktor „Dienstleistung" gering eingeschätzt. Die Studie führte zu einer Überprüfung des Erscheinungsbildes mit dem Ziel, Geschäfts- und Privatreisen als gleichrangige Faktoren zu kommunizieren. Und dies genau war der eigentliche Grund für die Beauftragung der Zürcher Agentur Zintzmeyer & Lux, der der Vorstand der Lufthansa eine weitgehendere Berücksichtigung emotionaler Faktoren zutraute als dem „Hyperrationalisten" Otl Aicher.

Die auffälligste Änderung der Agentur für die Corporate Identity der Lufthansa war die Bevorzugung der Farbe Gelb, die ein erheblich höheres Kommunikationspotential hat,

made a decision and had it carried out. This is also evidenced in the result. It was indeed good and, in substance, survives to the present day, but it was not the optimum, even though an excellent design department was established."[6]

As early as 1961 the Lufthansa Executive Board hired Hans G. Conrad (who as of 1953 had been one of the first students at the Design College at Ulm) as advertising director. Prior to that he had been the head of graphic design at Braun. In 1962 he was decisive in Lufthansa's commissioning Otl Aicher and his team. Klaus Wille, the first designer to be taken on as a permanent member of staff, and Hermann Roth both came from Braun and were hired on by Hans G. Conrad. From 1970 to about 1985 Ernst-Theodor Tschoepke was advertising director. During this period he, in collaboration with Klaus Wille, made possible the consistent continuation of design activities at Lufthansa. Frank Cesarz, for many years the public relations director, also had great impact on enhancing Lufthansa's design profile.

In a second stage, Lufthansa's own design department reworked the corporate identity as of 1966/67 using employees who, as was previously mentioned, stemmed from the Aicher "school." Certain suggestions were so close to the "master's" way of thinking that some published reports incorrectly saw Aicher as the author. This internal working group now envisioned a brightly polished belly on the airplane and the tail unit in blue, with a blue emblem on a circular yellow area. The "LM3" group had developed more than 25 tail unit variations for this revamping alone and they by no means found Aicher's approval. The blue stripe with the yellow logotype on the ground vehicles slipped quite a goodly ways downward, making the vehicles more elegant in appearance. All the subsidiaries profited as well from the typographical rationalism which Aicher preferred and which was retained even when the color schemes were changed. The identity continued to orient upon Aicher's design system even though ever greater deviations appeared in the colors, the typography and the photographic style – especially in depictions appearing in city guides or on local posters. Additionally, the passenger guidance system developed by Aicher was installed as of 1970. Jochen Eisenbrand explained:

"In the 1960s the need for internationally intelligible passenger guidance systems grew in parallel to growth in international air traffic. Among the first airports for which comprehensive signage was developed for passenger guidance were Schiphol Airport at Amsterdam (Total Design, 1965) and the airport at Frankfurt am Main (Otl Aicher, 1970)."[7]

Übervater Otl Aicher continued to be associated with the company even after the Executive Board in 1979 rejected his proposal – "Analysis 80" – to freshen up the corporate image. Now the Helvetica typeface was used in an even heavier embodiment, Helvetica Demibold. Here, again, the typography remained the same for all the subsidiaries and holdings, this in the interest of consistent corporate design. The rules applied to all Lufthansa corporate stationery – everything from flight and tendering documents to timetables, embossed service cards the size of credit cards, and the inflight entertainment program guide for the music piped into headphones in the aircraft. As early as 1978 the Lufthansa design department carried out a "modification of the Lufthansa logotype, company emblem and corporate colors, yellow and blue, for use in painting vehicles and aircraft as well as for lettering at downtown offices, stations, exhibitions." Application was made to the RAL Standards Committee to adopt two new shades in the standard color chart. This application was approved with a goodly dose of German thoroughness.

"Pursuant to application submitted by the Deutsche Lufthansa AG, Cologne, two new shades are adopted in the RAL 840 HR color chart. These shades are a fundamental

da diese Farbe – im Gegensatz zu Blau – im Wettbewerbsumfeld so gut wie nicht benutzt wird. So erschien selbst das Logo nun gelb auf weiß. Am Beginn ihrer Tätigkeit hatte der damalige Vorstand der Schweizer Agentur folgendes „Briefing“ an die Hand gegeben:

„Das vor 26 Jahren von Otl Aicher entworfene und fortlaufend weiterentwickelte Erscheinungsbild bedarf einer kritischen Prüfung, ob und inwieweit es die Zielsetzung, das Unternehmen Lufthansa auch in Zukunft als große, moderne, zuverlässige und souveräne Qualitätsfluggesellschaft, made in Germany, zu profilieren, noch unterstützt und visuell eindeutige Vorstellungen über Lufthansa erzeugen kann. Es wird die Auffassung vertreten, dass ein grundsätzlich anderer Auftritt weder notwendig noch richtig wäre. Allerdings sollte das Erscheinungsbild insgesamt überarbeitet, modernisiert und ‚auf Kurs‘ gebracht werden, beginnend mit der seit langem zur Entscheidung anstehenden Flugzeugkennzeichnung.“ [10]

Dies war so vorsichtig-diplomatisch formuliert, dass die Agentur Zintzmeyer & Lux sich entsprechend schwer tat. Zunächst wurden durch die Agentur Identity-Workshops veranstaltet, in denen die Persönlichkeitsmerkmale des Unternehmens verdeutlicht und konturiert wurden. Erst dann wurde die Farbgebung als Konsequenz aus ihrer in den Workshops erarbeiteten strategischen Positionierbarkeit überarbeitet. Eine Kombination von Grau, Blau, Gelb und Silber war schließlich die Lösung. Das Gelb wurde priorisiert, durch Silber und Grau anmutungsqualitativ gestützt. Das Blau wurde reduziert, aber bewusst für die Bildmarke eingesetzt. Die Agentur versuchte zunächst, bei der Neugestaltung der Flugzeugaußenkennzeichnung den Rumpf zu bereinigen. Da der zunächst favorisierte blanke Rumpf nicht realisierbar war, bot sich als Rumpffarbe Weiß an. In dieser Zeit flogen aber mehrere Gesellschaften, vor allem Konkurrenten in unmittelbarer Nähe (Air France, SAS) mit weißen Rümpfen. Zintzmeyer & Lux schlugen deshalb Rümpfe mit gelber Unterseite vor, die im Unternehmen und vor allem bei den Piloten jedoch schnell als „Gelbbauchunke“ verschrien waren. Man einigte sich schließlich auf eine Lösung mit grauer Rumpfseite. Die Leitwerkkennzeichnung setzte die Lufthansa-Tradition fort durch Bewahren des Lufthansa-typischen Bildes: So schlug Zintzmeyer & Lux die ursprüngliche Version „gelbes Leitwerk mit blauem Originalzeichen“ vor. Die bisherige Kranichscheibe wurde so geändert, dass das Firmenzeichen in seiner Originalform erscheint und damit als Kranich im Kreis besser wahrnehmbar wird: Aber an dieser Änderung war auch die interne LH-Designabteilung beteiligt. Bei der Fahrzeugkennzeichnung wurde die gelbe Volllackierung zurückgenommen. Als Farbträger wurde der be-

Vorschlag Zintzmeyer & Lux: Airbus A310-300 mit silbernem „Helvetica Demibold“-Schriftzug, gelbem Leitwerk und gelber Unterseite, 1988 (realisiert als Probelackierung)
Proposal by Zintzmeyer & Lux: Airbus A310-300 with silver-colored “Helvetica Demibold” logotype, yellow tail unit and yellow underbody, 1988 (test painting only)

element in the familiar Lufthansa symbol, which is used uniformly around the world. Appropriate reference colors are required for this purpose so as to ensure the most uniform possible coloration even where differing manufacturers make deliveries. The use of these colors extends to all areas of visual communications, such as aircraft and vehicle painting, external paint, plastic processing, dyeing textiles and printing on textiles, printed matter etc. Added to the application areas cited for Lufthansa are, moreover, the company's subsidiaries and holdings, such as Condor, Hansa Luftbild, Lufthansa Service GmbH. Account was taken of the fact that no comparable shades were previously present in the color chart and that the differences to similar shades already included in the chart are considerable. The new colors have been assigned the following designations and auxiliary descriptive terms: RAL 1028 melon yellow, RAL 5022 midnight blue." [8]

An article in the professional magazine *Design Report* mentioned in 1989, looking back at Otl Aicher's "Analysis 80":

"A decade ago Lufthansa executives began to re-think its public image. The intention was to revise the old livery crafted in Ulm, an image that in the meantime has become a part of design history. Otl Aicher had conceived it at the beginning of the 1960s with Hans 'Nick' Roericht at the College of Design at Ulm: a blue stripe and yellow tail unit with the blue crane. The fuselage is silver below and white above. After more than twenty years Lufthansa felt this to be too cool, too austere and too aristocratic. It was no longer high-tech but rather high-touch that was in demand. The aim was to speak to new customers at the emotional level. Thus Otl Aicher – at the beginning of the 1980s – was entrusted with reworking his own designs from the sixties. But Aicher wanted, at the very utmost, a revision on his own terms. This appeared to the airline to be in fact too cautious. That is why Lufthansa in 1986 charged a Zurich-based corporate identity agency, Zintzmeyer & Lux, with finding concepts for a new overall image." [9]

Apparently, however, events as they actually transpire are not always identical with public reporting. Already at the mid-1980s the Lufthansa Executive Board awarded to the American CRC – "Communication Research Center" – a contract for a study entitled "Image 2000." Referenced to the airline's corporate identity, the study revealed a deficit in emotional appeal and ascertained Lufthansa's position as an airline exclusively for business travelers, associated with high prices and an elitist image. In comparison, the company was given poor marks for service. The study resulted in a re-examination of the corporate identity with the goal of communicating the fact that

Designabteilung der Lufthansa: Airbus A320 mit „Helvetica Demibold"-Schriftzug und verstärkter Bildmarke, ab 1989 bis heute
Lufthansa design department: Airbus A320 with "Helvetica Demibold" and emphasized emblem, from 1989 to the present

währte Streifen beibehalten. Im weiteren Anwendungsbereich blieb Gelb die Primärfarbe. In zahlreichen Studien spielten die LH-Hausdesigner Farbkonzepte für die Flugzeugkennzeichnung durch. Da bekam das Leitwerk auch schon einmal schwarz-rot-goldene Landesfarben oder eine „stürzende" Wortmarke, und fast immer fiel das klassische Lufthansa-Blau so gut wie vollständig weg. Generell wurde die Bildmarke durch einen markanteren Kreis verstärkt und die Wortmarke aus Buchstaben des eigens entwickelten, fetteren Helvetica-Schnittes „Demibold" (ab 1978) konstruiert, die später „Lufthansa Corporate" genannt wurde und bis heute unverändert angewendet wird. Die von Zintzmeyer & Lux favorisierte Maschinenkennzeichnung, eine Bemalung mit gelbem Unterteil, weißem Oberteil und silberfarbenem Schriftzug verzichtete auf Otl Aichers kühles Blau. Ebenso erging es den Bodenfahrzeugen. Die Mitarbeiter der Fluggesellschaft fühlten sich düpiert: Vor allem die Angestellten äußerten ihren Unmut in der Firmenzeitschrift *Lufthanseat* in Leserbriefen. „Nicht unser Flugzeug" hieß es oder „zu nahe an Condor" und ein Mitarbeiter schrieb: „Man denke sich doch einmal aus, die Bundesrepublik krempelt ihre Nationalflagge um, sucht sich andere Farben und Motive: Die Identität wäre in Frage gestellt, ein Volksentscheid die Folge." [11] Ein „Farbenkrieg" also war intern im Unternehmen ausgebrochen und wurde durchaus öffentlich ausgetragen, so dass die Fachpresse Überschriften formulierte wie „Kranich auf Kollisionskurs" oder „Irrflug?". Hans „Nick" Roericht, der allein bzw. mit Aicher gemeinsam mehrere Projekte für die Lufthansa realisiert hat, hielt die Empörung für einen „guten Jux" und ein „dolles Element der jüngeren Firmengeschichte".

Der damalige Projektleiter „Corporate Identity" bei der Lufthansa, Wolfgang Weinert, stellte allerdings klar:

„Alles, was in Zürich geschehen ist, war immer eine Gemeinschaftsarbeit zwischen der Agentur und Lufthanseaten. Dieses gelb-weiße Designkonzept, mit dem wir im August durchs Land geflogen sind, ist auch das Ergebnis einer solchen Gemeinschaftsarbeit gewesen. Diesem Konzept lagen bestimmte Prämissen und Annahmen zugrunde, von denen sich nun herausgestellt hat, dass sie nicht mehrheitsfähig sind. Nicht mehrheitsfähig vor allem bei unseren Mitarbeitern, rund 87 % der Befragten haben es abgelehnt und etwa zwei Drittel der Kunden dachten ebenso. … Anhand einer Checkliste hat die Agentur Zintzmeyer & Lux vollständig alle denkbaren Kombinationen durchgespielt. Jeweils mit den Gestaltungselementen Rumpf, Rumpfunterseite, Triebwerke und Seitenruder und im Wechsel die Farben Blau, Weiß, Gelb und Silber. Dann ist man beim Befolgen des Grundsatzes ‚mit Gelb unterscheiden wir uns von den anderen' mit Gelb sehr weit gegangen – zu weit, wie wir nun wissen." [12]

Die Zürcher Agentur favorisierte also das Gelb als Erkennungsfarbe der Lufthansa und wollte dies auch bei den Flugzeugen verdeutlichen, obwohl die Bildmarke nach wie vor den blauen Vogel auf gelber Fläche zeigte. Deshalb schlugen sie für den Rumpf, die Leit- und Triebwerke einen melonengelben Anstrich vor, der allerdings auf den energischen Widerstand der „Vereinigung Cockpit" stieß. Eine gewisse Ähnlichkeit zu den von Otl Aicher in seiner „Analyse 80" entwickelten Ideen war unverkennbar. Das blieb auch Aicher selbst nicht verborgen, der daraufhin im August 1988 in der von ihm zeitlebens favorisierten Kleinschreibung an den Vorstand schrieb:

„sehr geehrter herr ruhnau, in einer veröffentlichung der *süddeutschen zeitung* vom 5. August dieses jahres werden die grundzüge eines neuen erscheinungsbildes der lufthansa vorgestellt. diese vorschläge gehen in sechs von sieben punkten auf ein konzept zurück, das ich seinerzeit dem vorstand der lufthansa vorgetragen habe. (kranich beibehalten, hausfarben gelb/blau, schriftzug in der helvetica, dominierende unternehmensfarbe gelb, nur schriftzug und zeichen in blau, heck gelb …" [13]

business and private travel were sectors of equal standing. And it was precisely this which was the real reason for commissioning Zintzmeyer & Lux in Zurich. The Lufthansa Executive Board felt that this agency would be better able to give greater consideration to emotional factors than could the "hyper-rationalist" Otl Aicher. The agency's most obvious change in the Lufthansa corporate identity was the increased use of the color yellow with its far greater communications impact, given the fact that, as opposed to blue, yellow is used hardly at all in the competitive environment. Thus even the logo itself then appeared in yellow on white. At the commencement of these activities the Executive Board then in office provided the Swiss agency with the following briefing:

"The corporate identity concept devised 26 years ago by Otl Aicher and continuously refined in the meantime requires a critical examination of whether and to what extent it still supports the goal of enhancing Lufthansa's profile in the future, marking it as a large, modern, reliable and authoritative air carrier distinguished by quality 'Made in Germany' and whether or not it can generate visually unequivocal images for Lufthansa. It is generally accepted that a complete change in identity is neither necessary nor would it be correct. The corporate identity should, however, be revamped throughout, modernized and be put 'on course,' beginning with aircraft markings, a subject which has been pending decision for some time now." [10]

This was worded so carefully and so diplomatically that the Zintzmeyer & Lux agency had its problems with it. The agency first held "identity workshops" where the features of the corporation's personality were made clear and given sharper contours. Only then was the color scheme reworked as the consequence of possibilities for strategic positioning as determined in the workshops. A combination of gray, blue, yellow and silver was finally identified as the solution. Yellow was assigned priority and the atmosphere it created was supported with silver and gray. The use of blue was reduced but quite consciously used for the emblem. The agency initially attempted to "tidy up" the fuselage when re-designing the exterior aircraft markings. Since the unpainted fuselage initially favored could not be realized, white offered itself as an alternate. At that time, however, several other airlines and above all some European competitors (Air France, SAS) were flying with a white fuselage. Zintzmeyer & Lux thus suggested aircraft with a yellow lower section. Inside the company and above all among pilots this idea soon fell into disrepute and it came to be called the "yellow-bellied toad." Agreement was finally reached with a solution featuring a gray underbelly. The markings on the tail unit continued Lufthansa tradition by retaining the look so typical of Lufthansa. Consequently Zintzmeyer & Lux proposed returning to the original version: "... yellow tail unit with the original symbol in blue." The crane previously used was modified so that the corporate emblem appeared in its original form and thus was more easily perceived as a crane within a circle. The internal Lufthansa design department was also involved in this change. As regards ground vehicle markings, the all-over yellow painting scheme was rescinded. The time-tested stripes were retained as a color element and in other application areas yellow remained the dominant color.

In the course of numerous studies Lufthansa's own designers experimented with color concepts for the aircraft markings. In one case the rudder was adorned with Germany's national colors, black, red and gold, or with a "tumbling" logotype, and in almost all cases the traditional Lufthansa blue disappeared almost completely. In general the emblem was strengthened with a more prominent circle while the logotype was made up of the letters drawn from a specially developed and heavier Helvetica "Demibold" family (as of 1978), which was later named "Lufthansa Corporate" and is still in use, unchanged, today. The aircraft livery favored by Zintzmeyer & Lux – comprising a

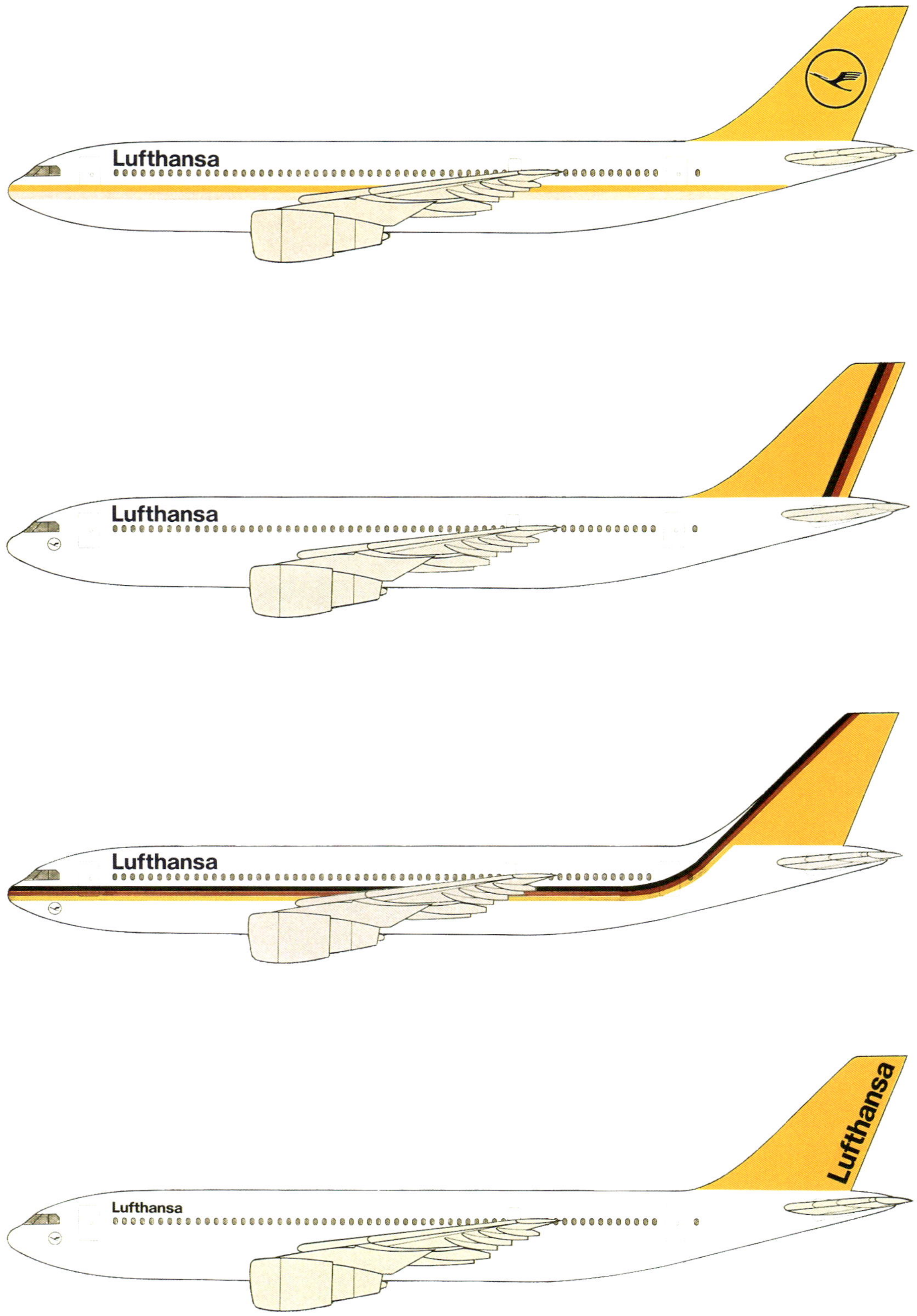


Designabteilung der Lufthansa: Studien für Flugzeugbemalung eines Airbus A300, um 1988
Lufthansa design department: Studies for the Airbus A300 livery, about 1988

yellow underbody, white upper section and silver-colored logotype – did entirely without Otl Aicher's cool blue tone. The ground vehicles underwent a similar change. The airline's employees felt that they had been duped. They expressed their displeasure above all in letters to the editor of the Lufthanseat company magazine. "It's not our airplane," they wrote, or "too similar to Condor." One associate even penned, "Just imagine what would happen if Germany were to revamp the nation's flag and were to search for different colors and motifs: national identity would be shaken to the core and a public referendum the consequence." [11] A "war of the colors" had thus broken out within the company and was fought in full public view so that the trade press published headlines such as "Crane on a crash course?" or "Adrift in the fog?" Hans "Nick" Roericht, who either alone or together with Aicher had finalized numerous projects for Lufthansa, felt the uproar was a "good joke" and a "droll episode in the company's recent history." The Lufthansa "Corporate Identity" project manager at the time, Wolfgang Weinert, set the record straight:

"Everything that happened in Zurich was always a joint effort between the agency and Lufthansa staff. This yellow and white design concept, with which we flew around the country in August, was also the result of such a combined effort. This concept was based on certain premises and presumptions which, as it turned out, were not capable of finding majority approval. And that majority was lacking above all among our employees; about 87 per cent of those queried rejected the concept and about two-thirds of the customers thought the same way … Using a checklist as the basis, Zintzmeyer & Lux worked their way through every conceivable combination. In each case it considered the design surfaces – fuselage, belly, engine cowlings and rudders – and alternated the colors blue, white, yellow and silver. Then, in following the principle of 'with yellow we distinguish ourselves from the rest,' we pushed yellow hard – too hard, as we now know." [12]

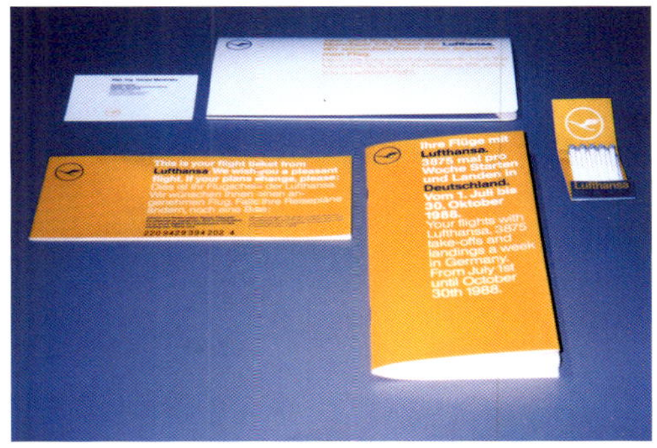

Studie Zintzmeyer & Lux (nicht realisiert): Erscheinungsbild Printmedien, Ende 1980er Jahre
Study by Zintzmeyer & Lux (not implemented): Design guidelines for print media, late 1980s

The Zurich agency thus favored yellow as the Lufthansa recognition color and also wanted to make this clearer on the aircraft, even though the emblem continued to show the blue crane on a yellow surface. That is why they suggested a melon yellow coat of paint for the fuselage, the tail unit and the engine cowlings which, however, met with energetic resistance from the Vereinigung Cockpit – the German pilots' association. A certain similarity with the ideas developed by Otl Aicher in his "Analysis 80" could not be overlooked. Neither was this lost on Aicher himself. He wrote to the Executive Board in August of 1988:

"Dear Mr. Ruhnau: The fundamentals of a new Lufthansa corporate identity were unveiled in an article published in the *Süddeutsche Zeitung* on August 5 of this year. In six of seven items, these proposals hark back to a concept which I earlier presented to the Lufthansa Executive Board (retain crane, yellow and blue corporate colors, Helvetica logotype, yellow as the dominant corporate color, only the lettering and symbols in blue, tail unit yellow …" [13]

Wolfgang Weinert did not see Aicher's astonishment, expressed as it was, as being a problem. In response to the question, "Had any thought been given to continuing to work with the Aicher organization?" he answered diplomatically: "You shouldn't forget that our own in-house designers are, as it were, working as Aicher's students in our company." [14] And thus it came to pass that at the beginning of 1990 the Lufthansa Design Center operating at that time (1989–92) under the management of Dr. Bernd Weidemann and Christian Steguweit, together with the Design Department at Cologne, managed by Renate Liane Schneider, were charged with working out a forward-looking concept for the Lufthansa identity in the 1990s and to document this in design manuals.

Aichers solchermaßen ausgedrückte Verwunderung allerdings sah Wolfgang Weinert nicht als Problem. Auf die Frage: „Hat es Überlegungen gegeben, mit dem Büro Aicher weiterzuarbeiten?", antwortete er diplomatisch: „Sie dürfen nicht vergessen, dass unsere internen Designer gewissermaßen als Schüler von Aicher in unserem Unternehmen tätig sind."[14] Und so wurde dann Anfang des Jahres 1990 das damalige Lufthansa Design Center (1989–92) unter der Leitung von Dr. Bernd Weidemann und Christian Steguweit zusammen mit der Kölner Designabteilung unter der Leitung von Renate Liane Schneider in Köln beauftragt, ein längerfristiges Konzept für den Lufthansa-Auftritt der 1990er Jahre zu erarbeiten und in Design Manuals zu dokumentieren.

Insgesamt sind die in Zusammenarbeit mit der Agentur Zintzmeyer & Lux erfolgten Aktualisierungen des Erscheinungsbildes deutlich, aber einen grundsätzlichen Imagewechsel bedeuten sie nicht. Ihre generelle Strategie mag man als „fortschreibende Kontinuität" begreifen.

Pentagram: Logo Star Alliance zur Kennzeichnung der Mitglieder, 1997
Pentagram: Star Alliance logo to identify the members, 1997

Otto Firle, jenem Grafiker, der 1918 das Kranich-Logo kreierte, wäre vermutlich schon dies zuviel gewesen. In einem Gedächtnisprotokoll eines Gespräches von 1965 zwischen dem bereits hochbetagten Entwerfer und einem Herrn Rahskopff, das in dem Firmenarchiv der Lufthansa in Köln dokumentiert ist, heißt es:

„In Berlin entwarf Firle das Firmenabzeichen, für das ihn weder ein Abzeichen einer französischen Fliegerstaffel noch eine Abbildung von einem Kranich oder einem Greif als Vorbild gedient hatte. Er wollte vielmehr in seiner Zeichnung das Fliegerische und das Technische, die im entstehenden Flugverkehr vereint sind, und deren kommende Entwicklung, also deren Aufwärtsbewegung, zum Ausdruck bringen. Die Farben für das Firmenzeichen wählte er in einem Tiefblau auf gelbem, mehr orangegelbem Untergrund. An einen Kreis oder etwas Ähnliches um das Firmenzeichen herum war überhaupt nicht gedacht. ‚Ich sperre ein aufsteigendes Firmensymbol doch nicht in einen Käfig; das wäre einfach absurd!', erklärte Prof. Firle."[15]

Die emblematische Qualität des Kreises um den Kranich herum, die dieses Signet ja überhaupt erst zur Bildmarke macht, ist dem Entwerfer offenbar dann doch entgangen.

Für das weltweite Transportnetzwerk Star Alliance, zu dem anfangs die Fluglinien Air Canada, Lufthansa, SAS, Thai Airways und United Airlines gehörten und das heute auf weit über ein Dutzend Fluglinien angewachsen ist, entwarf die Designagentur Pentagram die visuelle Identität. Die Erkennbarkeit der Allianz und die Selbstständigkeit der Mitglieder sollten gleichzeitig in einem Symbol erkennbar werden. Pentagram entwickelte ein Signet aus fünf Einzelelementen, die wie ein fünfzackiger Stern in verschiedene Richtungen weisen. Die fünf Elemente – räumlich dargestellte Tetraeder – bilden ihrerseits wieder einen Gesamtstern. Justus Oehler, einer der Pentagram-Partner in London, erläuterte: „Wir hatten die fünf verschiedenen Kulturen der Partnerfluglinien zu berücksichtigen und mussten sicherstellen, dass die Marke und der Name rund um die Welt funktionieren." Auch den Namen „Star Alliance" kreierte Pentagram. Ab Anfang 1997 sind Wort- und Bildmarke auf jedem Langstreckenflieger der LH in der Nähe des Cockpits Teil der Flugzeugkennzeichnung. Aber inzwischen sind weitere Mitglieder dem Netzwerk beigetreten. Insofern wurde das Logo ebenso von der Realität überholt wie die Fahne der Europäischen Gemeinschaft nach dem Beitritt neuer Mitgliedsstaaten.

Auch wenn sich das Erscheinungsbild der Lufthansa durch solche wirtschaftlich-strategischen Konzernentscheidungen möglicherweise weiter ausweiten wird, bedarf es wenig Prophetie, um vorherzusagen, dass die Basiselemente des Corporate Designs der Lufthansa – die Farben Gelb und Blau, ergänzt um Weiß und Grau, ebenso wie der Kranich – bleiben werden. Bei aller Kreativität der Agenturen wäre es kontraproduktiv,

All in all, the modernization of the corporate identity conceived by Zintzmeyer & Lux is quite clear, but it by no means represents a fundamental change in the corporate image. One might sense the general strategy to be one of "ongoing continuity."

This would presumably have been too much for Otto Firle, the graphic artist who created the crane logo in 1918. In the minutes of a 1965 discussion between the creator of the symbol, already well on in years, and a certain Mr. Rahskopff, documented in the Lufthansa Company Archives in Cologne, it says:

"In Berlin Firle drafted the company symbol, for which neither the badge of a French air force wing nor an illustration of a crane or a griffin had served as the basis. He wanted instead in his drawing to give expression to flight and technical expertise, joined in air travel then coming into being, and to express their upcoming developments and thus their upward movement. The colors he selected for the company logo were a deep blue on yellow, or more yellow-orange, background. He had not given any thought whatsoever to using a circle or anything similar around the company symbol. 'I wouldn't trap a rising company's emblem in a cage; that would be simply absurd!' Professor Firle declared." [15]

The emblematic quality of the circle around the crane, which in fact elevates this mark to the quality of a logo, apparently escaped the attention of its creator.

The Pentagram design agency developed the visual identity for the globe-spanning Star Alliance transportation network. The original members were Air Canada, Lufthansa, SAS, Thai Airways and United Airlines; today this group has grown to include more than a dozen air carriers. A new symbol was to be created – one which identified the Alliance itself while at the same time emphasizing the autonomy of the individual airlines. Pentagram developed a mark made up of five individual elements which, like a pentacle, all point in different directions. These five elements – tetrahedra shaded for a three-dimensional effect – act together to form a complete star. Justus Oehler, one of Pentagram's partners in London, explained: "We had to take account of the five different cultures embodied by the participating airlines and at the same time ensure that the trademark and the name would function all around the world." Pentagram also devised the "Star Alliance" name. As of the beginning of 1997 the emblem and logotype, positioned near the cockpit, have graced every Lufthansa long-haul aircraft. Additional members have joined the network in the meantime. And thus the logo has been overtaken by reality, sharing this fate with the flag of the European Union as new nations joined.

Even though the Lufthansa corporate identity might be further expanded by such business and strategic decisions, one hardly needs the gift of prophecy to predict that the basic elements in the corporate design – the colors yellow and blue, augmented with white and gray, and of course the crane – will remain. In sprite of all the creativity which the agencies might bring to bear, it would be counterproductive to jettison a long-standing identity in a single abrupt change.

One can, however, certainly update the identity and will surely have to do so. The Lufthansa subsidiaries, as well, will require a corporate design whenever they are founded or restructured. It must be independent but the relationship with the parent company should also be recognizable.

With the beginning of the new millennium the Lufthansa "brand" attained new significance within the entire field of strategic options. The brand became more important and corporate management paid more attention to modern brand management and care than it had been the case in earlier decades. It was a rare stroke of good luck that a strong brand was already available. Numerous projects were launched with the

Pentagram: Logo Star Alliance, 1997
Pentagram: Star Alliance logo, 1997

ein lange gewachsenes Erscheinungsbild mit einem harten Schnitt über Bord zu werfen. Wohl aber kann und muss man es aktualisieren und auch die Tochterunternehmen der Lufthansa, die neu strukturiert oder neu gegründet werden, bedürfen eines zwar selbstständigen Corporate Designs, das aber erkennbar zum Mutterunternehmen in Relation stehen sollte.

Mit dem Anfang des neuen Jahrhunderts bekam die „Marke“ Lufthansa eine neue Bedeutung im gesamten Feld der strategischen Optionen – sie wurde wichtiger, und ihre zeitgemäße Führung und Pflege rückte mehr als in vorherigen Dekaden in den Fokus der Unternehmungsführung. Der seltene Glücksfall, über eine starke Marke zu verfügen und diese mit neuen, zeitgemäßen Methoden und Praktiken zu entwickeln, wurde mit einer Vielzahl von Projekten in Angriff genommen.

Ab 2001 wurde die Peter Schmidt Group sowohl mit markenstrategischen Aufgaben als auch mit Entwurfsprojekten für die Lufthansa beauftragt. Zu den ersten gehörten eine Visualisierung des bestehenden Markenleitbildes, ein aktualisiertes Designleitbild sowie eine sogenannte „Markenarchitektur“. Ebenso gehört zu den strategischen Projekten ein Marketing-Leitfaden für alle regionalen Partner-Airlines. Die Peter Schmidt Group entwickelte zudem ein Flugzeugbranding für die erwähnten Partner-Airlines. Zum Designleitbild gehören die grundsätzliche Festlegung in den Bereichen Farben, Formen, Materialien und Oberflächen, aber etwa auch Überlegungen zur Besucherführung, zur Empfangssituation im Außen- und Innenbereich, zur Zonierung der Aufenthaltszonen, zur Beleuchtung und zu den Sanitärbereichen. Dabei sind die Designvorgaben in den verschiedenen Flugklassen durchaus unterschiedlich. So werden etwa bei den verwendeten Farben in der Economy Class die Grauwerte reduziert und die CI-Farben Gelb und Blau verstärkt eingesetzt. In der Business Class überwiegt die Farbe Blau und in der First Class kommen zu diesem Blau zunehmend warme Naturtöne hinzu. Ebenso unterscheiden sich die eingesetzten Materialien: In der Economy Class finden haptisch anregende Oberflächenstrukturen Verwendung, in der Business Class innovative Materialien und Oberflächen, in der First Class edle Naturmaterialien wie Hölzer und Leder. Man erkennt die Design-Strategie: Es geht nicht um fertige Gestaltungsvorschläge, sondern zunächst um segmentspezifische Anmutungen für verschiedene Passagierklassen und um gewissermaßen psychologische Profile von Kundenansprachen. Innerhalb einer solchen differenzierten Syntax bleibt dann genügend Spielraum für individuelle Entwürfe und Designlösungen auch von anderen Gestaltern. Insofern schränkt das Designleitbild Kreativität nicht ein, wohl aber wird sie behutsam kanalisiert – und ab und zu durchaus

ATB (Automated Ticketing Boarding)-Bordkarte, 1990er Jahre
Boarding pass for ATB (Automated Ticketing and Boarding), 1990s

Peter Schmidt Group: ATB (Automated Ticketing Boarding)-Bordkarte, ab 2004
Peter Schmidt Group: Boarding pass for ATB (Automated Ticketing and Boarding), as of 2004

goal of applying contemporary methods and practices to further develop the brand. Beginning in the year 2001 the Peter Schmidt Group was charged with carrying out both strategic, brand-related tasks and developing specific drafts for Lufthansa. Among the first assignments were a visualization of the existing brand guidelines, drafting an updated design guideline and defining the so-called "brand architecture." Also among the strategic projects is a marketing guideline for all the regional partner airlines. Among the elements considered in the design guidelines are fundamental specifications in regard to colors, shapes, materials and surfaces. Considered beyond that are visitor and passenger guidance, the landside and airside reception situations, zoning for the waiting areas, illumination and sanitary facilities. Here the design specifications for the various classes of travel do differ. So, for example, the use of gray in Economy Class is reduced and the two major CI colors, yellow and blue, are employed more prominently. Blue predominates in Business Class while in First Class this blue is complemented to an ever-increasing extent with warm, natural colors. The materials used differ in much the same way. Surface textures which are tactilely interesting predominate in Economy Class. In Business Class we find innovative materials and surfaces and in First Class more costly natural materials such as woods and leather. The design strategy is apparent. We are not working with cut-and-dried design proposals but instead with specific décors for various classes of travel and, to a certain extent, are employing psychological profiles when addressing customers. Even a highly differentiated syntax such as this leaves sufficient leeway for individual concepts and solutions created by other designers. To this extent, the design guidelines do not restrict creativity but rather gently channel it and certainly will use "illustrative décor sketches" from time to time to provide examples of space planning. One such design concept, created by the Peter Schmidt Group, was the basis for updated lounge and terminal planning aimed at top customers. This was turned into reality in December 2004 at the Frankfurt Airport. This concept was devised in cooperation with the "bhar" design office. Also included among the newly adopted methods was a so-called internal Brand Book. In its third edition the Chairman of the Lufthansa Executive Board, Wolfgang Mayrhuber, quite correctly emphasized:

Peter Schmidt Group: Das Design-Leitbild der Lufthansa für die First Class, ab 2004
Peter Schmidt Group: Lufthansa First Class design guidelines, as of 2004

auch mit „Anmutungsskizzen“ für beispielhafte Raumplanungen versehen. Ein solches Designkonzept der Peter Schmidt Group war die Grundlage für eine aktualisierte Lounge- und Terminalplanung für Top-Kunden, welche bis zum Dezember 2004 auf dem Frankfurter Flughafen realisiert wurde. Sie entstand in Zusammenarbeit mit dem Designbüro „bhar“. Weiterhin gehörte zu diesen neuen Methoden auch ein sog. internes „Markenbuch“, in dessen dritter Auflage der Vorstandsvorsitzende der Lufthansa, Wolfgang Mayrhuber, zu Recht betont:

„So wie das Angebot der Lufthansa mit den Jahren vielfältiger und komplexer geworden ist, so ist auch die Marke vielschichtiger geworden. Sie hat sich im Laufe der Zeit wie eine Persönlichkeit entwickelt, hat bestimmte Eigenschaften angenommen und sich mit einem differenzierten Leistungsbild profiliert. Die Marke Lufthansa ist dabei zu einem der wichtigsten materiellen und ideellen Werte der Fluggesellschaft geworden. Die Bewahrung und die weitere Steigerung dieses Wertes ist daher ein vorrangiges Ziel für Management und Mitarbeiter. Eine entscheidende Voraussetzung, um dieses Ziel zu erreichen, ist die Schaffung eines Instruments, das die strategische Ausrichtung transparent macht und konkrete Hilfestellung für die Umsetzung bietet. Dazu gehören elementare Regeln zur Führung der Marken sowie deren Visualisierung im Corporate Design – verbindlich, konsistent und langfristig festgelegt und für einen breiten Nutzerkreis verfügbar.

Das Markenbuch soll diese Funktion erfüllen. Es schafft ein gemeinsames Markenverständnis und gibt Orientierungshilfe für die tägliche Arbeit, in erster Linie als Instrument für die Planung, Umsetzung und Kontrolle von markenrelevanten Maßnahmen.“ [16]

Innerhalb der Markenführung spricht man dann von „Markenarchitektur“, wenn Produkte und Serviceleistungen in einer Hierarchie eingegliedert werden und Ordnungsprinzipien zur Benennung und Gestaltung dieser Produkte mit Serviceleistungen geschaf-

Peter Schmidt Group: Flugzeug „Lufthansa Regional“ mit „Helvetica“-Schriftzug und neuer Leitwerkbemalung, ab 2003
Peter Schmidt Group: “Lufthansa Regional” aircraft with “Helvetica” logotype and new tail unit design, as of 2003

"Just as Lufthansa's offerings have become more diverse and complex over the years, the brand has also become more multi-layered. Over the course of time, the brand has developed, like a personality, assumed specific characteristics and differentiated its performance image. The Lufthansa brand has become one of the airline's most important material and intangible assets. Therefore preserving and increasing its value is a paramount goal for management and employees. A critical prerequisite to achieving this goal is creating an instrument to make the strategic focus transparent and to offer concrete assistance in implementation. This includes elementary rules concerning the management of the brand and its visualization in corporate design defined in a binding, consistent and lasting manner and making it available to a broad group of users. The Brand Book is to fulfill this function. It creates a common brand understanding and gives orientation aid for day-to-day use, primarily as an instrument for planning, implementation and control of brand-relevant measures." [16]

Within the general field of brand management one refers to "brand architecture" whenever products and services are to be organized hierarchically and principles are set down for the nomenclature and design of these products and services. The objectives behind the "brand architecture" are thus four-fold: to increase the value of the brand, to pave the way for more efficient decision making with the support of brand architecture matrices and specifications as regards corporate, internal, public and general communications design, to install a carefully controlled sequence of definition processes for both design and content, and to create a consistent overall identity for the corporation both inside and outside the organization.

The "Lufthansa Regional" subsidiary flies within Germany and on European routes. The Peter Schmidt Group developed a new tail unit logo for this brand, its colors referencing the blue and yellow CI colors. It is made up of a dynamic matrix of dots and thus, in abstract form, symbolizes the European route system.

All these strategic considerations, drawn together in the Brand Book, emphasize the ambivalence between functional and emotional brand values. Quite apparent, however, is that Lufthansa is moving away from its image as a high-tech airline and toward that of a high-touch carrier.

In a manual on corporate identity, under the heading "Integrated communication and corporate design at Lufthansa today," it is noted:

"The function of corporate design at Lufthansa has grown. In addition to ensuring uniformity, visualizing the brand values is becoming more important. Operating on fixed principles, the communication image hovers between creativity and consistency. Design guidelines, documented in a system of several manuals, provide information on both constants and variables. Detailed instructions on the use of color must do justice both to a restrained corporate image in business communications and to the prominent use of color for differentiation in the competitive environment. In the fields of advertising and sales promotion it is necessary to integrate creative ideas, in line with current market expectations, into a layout which is typical for Lufthansa. The aim here is to guarantee quick attention-getting and to ensure recognition. The market and the corporation are ever more often demanding modifications of the design at short notice. That is why Lufthansa has, since 2001, made major portions of the existing manuals available online. Using the intranet, employees all around the world can access current information and materials. Outside agencies use password-protected Internet portals to access the website. Available to the worldwide network of agencies is a dedicated online instrument for exchanging data and special copy or artwork for specific markets, arranged by subjects. Authorized use and updating of photo and image materials is controlled by way of a

1 Keith Lovegrove, *Airline. Identity, Design and Culture*, New York 2000, S. 109, 126 f.
2 Jochen Eisenbrand, Erläuterungstext in der Ausstellung „Airworld", Vitra Design Museum, Weil am Rhein 2004.
3 Vortrag „Flugzeug- und Fahrzeugkennzeichnung", September 1988, Lufthansa-Archiv, Frankfurt am Main.
4 Bernd Franck, „Und wie man sich zeigt, so ist man. Rückblick. Vor 35 Jahren verpasste Otl Aicher der Lufthansa ein neues Erscheinungsbild", *form spezial*, Nr. 2, 1988, S. 68 f.
5 Bernd Franck, a. a. O., S. 69.
6 Otl Aicher, zit. nach Bernd Franck, a. a. O.
7 Jochen Eisenbrand, a. a. O.
8 Neue RAL-Farben, Briefdokument, August 1979, Lufthansa-Archiv, Frankfurt am Main.
9 Klaus Thomas Edelmann, „Irrflug?", *Design Report*, Nr. 9, 1989, S. 12.
10 Vortrag „Flugzeug- und Fahrzeugkennzeichnung", September 1988, Lufthansa-Archiv, Frankfurt am Main.
11 Zit. nach Klaus Thomas Edelmann, a. a. O.
12 „Das gelb-weiße Designkonzept ist durchgeflogen", *Design Report*, Nr. 9, 1989, S. 13 ff.
13 Otl Aicher, zit. nach Bernd Franck, a. a. O., S. 69.
14 Zit. nach: *Design Report*, Nr. 9, 1989, S. 15.
15 Gedächtnisprotokoll von Hermann Rahskopff nach einem Besuch bei Otto Firle in Düsseldorf am 18. 11. 1965, Lufthansa-Archiv, Frankfurt am Main.
16 Wolfgang Mayrhuber, Vorwort, in: *Lufthansa-Markenbuch 3*, Deutsche Lufthansa AG, Frankfurt am Main 2004.
17 „Lufthansa – die konsistente Entwicklung einer Corporate Identity", in: Klaus Birkigt/Marinus M. Stadler/Hans J. Funck (Hrsg.), *„Corporate Identity" – Grundlagen, Funktionen, Fallbeispiele*, Redline Wirtschaft, bei Verlag Moderne Industrie, Frankfurt am Main 2002, S. 438, 439.

fen werden. Die Ziele einer „Markenarchitektur" bestehen also in einer Erhöhung des Markenwertes, im Bereitstellen effizienter Entscheidungsfindungen durch markenarchitektonische Raster und Vorgaben in Corporate-, Interior-, Public- und Kommunikationsdesign; in einem kontrollierten Ablauf von Design- und inhaltlichen Definitionsprozessen sowie in einem schlüssigen Gesamteindruck des Unternehmens nach innen und außen. Die Lufthansa-Tochter „Lufthansa Regional" fliegt auf innerdeutschen und europäischen Routen. Für diese Marke entwickelte die Peter Schmidt Group ein neues Leitwerklogo als Flugzeugaußenkennzeichnung, das sich farblich auf die CI-Farben Blau und Gelb bezieht, aus einem dynamischen Punktraster besteht und so in abstrahierter Form das europäische Streckennetz symbolisiert.

All diese strategischen Überlegungen, zusammengefasst im „Markenbuch", betonen die Ambivalenz zwischen funktionalen und emotionalen Markenwerten. Deutlich wird, dass sich die Deutsche Lufthansa von dem Image einer High-Tech-Fluggesellschaft hin zu einer High-Touch-Fluggesellschaft bewegt.

Unter der Überschrift „Integrierte Kommunikation und Corporate Design der Lufthansa heute" heißt es in einem Fachbuch zur „Corporate Identity":

„Die Aufgabe des Corporate Designs bei Lufthansa ist gewachsen. Neben der Sicherstellung der Einheitlichkeit geht es zunehmend um die Visualisierung der Markenwerte. Auf Basis fester Grundregeln bewegt sich der Kommunikationsauftritt zwischen Kreativität und Konsistenz. In einem mehrteiligen Manualsystem dokumentierte Gestaltungsrichtlinien geben Auskunft über Konstanten und Variablen. Ein detailliertes Farbverhalten muss sowohl dem zurückhaltenden Unternehmensauftritt im Bereich der Bürokommunikation als auch dem dominanten Farbeinsatz zur Differenzierung im Wettbewerb gerecht werden. In der Werbung und Verkaufsförderung gilt es, marktgerechte Kreativideen in ein Lufthansa-typisches Layout zu integrieren, um schnelle Aufmerksamkeit, aber auch Wiedererkennbarkeit sicher zu gewährleisten. Markt und Unternehmen erfordern immer kurzfristigere Anpassungen des Designs. Darum bereitet Lufthansa seit 2001 wesentliche Teile des bestehenden Manualsystems „online" auf. Über das Intranet können Mitarbeiter weltweit auf die aktuell aufbereiteten Informationen und Materialien zugreifen. Externe Agenturen gelangen passwortgeschützt über Internet auf die Website. Dem weltweiten Agenturnetzwerk steht ein eigenes Online-Instrument zum Austausch der Daten und speziellen Vorlagen für Märkte nach Themen geordnet zur Verfügung. Über eine zentrale Bilddatenbank wird die rechtmäßige Nutzung und Aktualisierung des Bildmaterials gesteuert. Mitarbeiter marken- und designrelevanter Bereiche werden über Workshops und Broschüren mit den Systemen vertraut gemacht. Neue Produkte werden von den Fachbereichen zentral auf Basis der grundlegenden Marken- und Designwerte erarbeitet." [17]

Und so werden sich die Corporate Identity, das Branding, die Designleitbilder und die „Markenarchitektur" weiterentwickeln. Viele sind beteiligt: Architekten und Produktdesigner, Kommunikationsdesigner und Public-Design-Entwerfer, Werbeagenturen und Branding-Agenturen. All deren Aktivitäten bündelt im Unternehmen die hauseigene Corporate-Design-Abteilung, zu der auch das „Markenmanagement" gehört. Sie ist und fungiert als das „gestalterische Gewissen" der Lufthansa.

central images database. Employees in areas which are relevant to branding and design are made familiar with the system through workshops and brochures. New products are created centrally, by the various departments concerned, on the basis of the fundamental brand and design values."[17]

And thus corporate identity, branding, design guidelines and "brand architecture" will continue to be developed. Many professions are involved here: architects and industrial designers, communications designers and public design specialists, advertising offices and branding agencies. All these activities are drawn together within the company in its own Corporate Design Department, since "brand management" is under its purview. It is and will continue to function as the "design conscience" at Lufthansa.

**1** Keith Lovegrove, *Airline. Identity, Design and Culture*, New York, 2000, pp. 109, 126 f.
**2** Jochen Eisenbrand, Explanatory copy for the "Airworld" exhibit, Vitra Design Museum, Weil am Rhein, 2004.
**3** Speech, "Aircraft and vehicle livery," September 1988, Lufthansa Archives, Frankfurt am Main.
**4** Bernd Franck, "Und wie man sich zeigt, so ist man. Rückblick. Vor 35 Jahren verpasste Otl Aicher der Lufthansa ein neues Erscheinungsbild," *form spezial*, no. 2, 1998, pp. 68 f.
**5** Bernd Franck, op. cit., p. 69.
**6** Otl Aicher, quoted by Bernd Franck, op. cit.
**7** Jochen Eisenbrand, op. cit.
**8** New RAL shades, letter dated August 1979, Lufthansa Archives, Frankfurt am Main.
**9** Klaus Thomas Edelmann, "Irrflug?", *Design Report*, no. 9, 1989, p. 12.
**10** Speech, "Aircraft and vehicle livery," September 1988, Lufthansa Archives, Frankfurt am Main.
**11** Quoted after Klaus Thomas Edelmann, op. cit.
**12** "Das gelb-weiße Designkonzept ist durchgeflogen," *Design Report*, no. 9, 1989, pp. 13 ff.
**13** Otl Aicher, quoted by Bernd Franck, op. cit., p. 69.
**14** Quoted in *Design Report*, no. 9, 1989, p. 15.
**15** Minutes by Hermann Rahskopff, after a visit to Otto Firle in Düsseldorf on November 18, 1965, Lufthansa Archive, Frankfurt am Main.
**16** Wolfgang Mayrhuber, "Preface," in: *Lufthansa Markenbuch 3*. Deutsche Lufthansa AG, Frankfurt am Main, 2004.
**17** "Lufthansa – die konsistente Entwicklung einer Corporate Identity," in: Birkigt/Stadler/Funck (ed.), *Corporate Identity – Grundlagen, Funktionen, Fallbeispiele*, Redline Wirtschaft series, published by Verlag Moderne Industrie, Frankfurt am Main, 2002, pp. 433, 439.

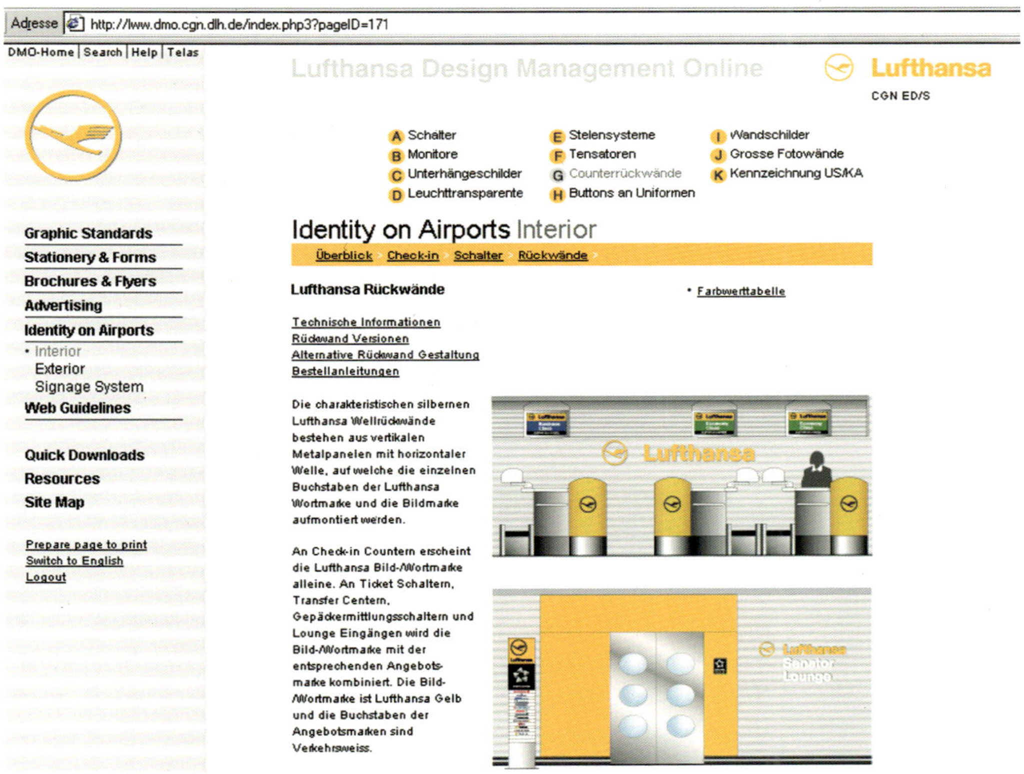


True Brand (ehem. Luxon Carra): Design Management Online, ab 2000
True Brand (formerly Luxon Carra): Design Management Online, as of 2000

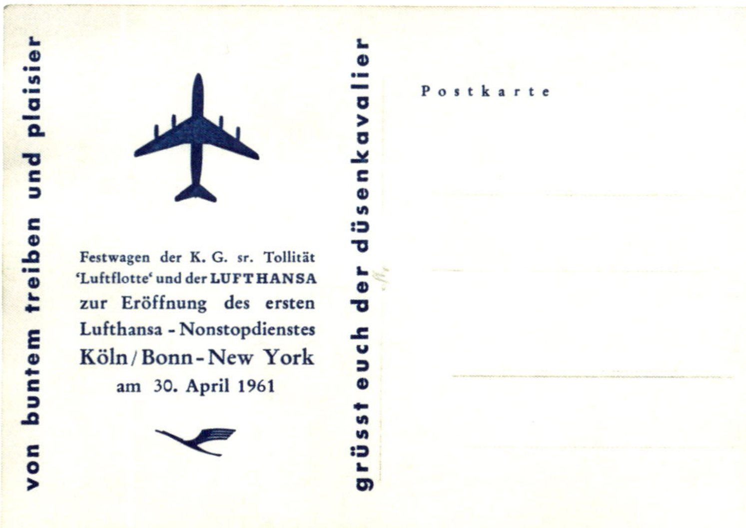

Vorder- und Rückseite Werbepostkarte „Festwagen der Tollität Luftflotte", 1961
Promotional postcard, front and back: Carnival float depicting the jolly air fleet, 1961

Notizblock, um 1960
Note pad, about 1960

Deutsche Lufthansa: Jahresbericht 1955, 1957, 1960
Deutsche Lufthansa: Annual Reports, 1955, 1957, 1960

Bordkarten-Hülle mit der Lockheed Super Star Constellation, Ende 1950er Jahre
Boarding pass envelope showing the Lockheed Super Star Constellation, late 1950s

Kalendertitelblatt, 1957
Calendar cover sheet, 1957

Werbebroschüre, um 1960
Advertising brochure, about 1960

Gepäckanhänger, Vorder- und Rückseite, um 1960
Luggage tag, front and rear, about 1960

Ticketcover, Gepäckschein und -anhänger, um 1960
Ticket envelope, baggage claim check and luggage tag, about 1960

Zigarrettenbanderolen, um 1960
Cigarettes bands, about 1960

Flugzeug-Schlepper, Ende 1950er Jahre
Airplane tug, late 1950s

Traktor und Gepäckwagen, Ende 1950er Jahre
Tractor and luggage van, late 1950s

Transportfahrzeug, Ende 1950er Jahre
Van, late 1950s

Fahrrad für Transporte auf dem Flughafengelände, Ende 1950er Jahre
Bicycle for transportation on the airport grounds, late 1950s

VW „Bully" für Personentransport, um 1960
VW van for passenger transportation, about 1960

**Lufthansa**

Deutsche Lufthansa AG
5 Köln Claudiusstraße 1

Mitglied der IATA

Telefon: Durchwahl-Nummer
des Sachbearbeiters (0221) 2068 . . .
Vermittlung: (0221) 20681
Telegramme: Lufthansa Köln
Telex: 08882902
Banken: Deutsche Bank AG, Köln, 106 141
Rheinische Girozentrale und Provinzialbank
Köln, Konto 4133
Dresdner Bank AG, Köln, Konto 14-122270
Commerzbank AG, Köln, Konto 13123
Bank für Gemeinwirtschaft AG, Köln, 17170
Landeszentralbank, Köln, Konto 3/72
Postscheck: Köln, Konto 414

| Ihre Zeichen | Ihre Nachricht vom | Unsere Zeichen (Bitte wiederholen) | Datum |
|---|---|---|---|

Vorsitzender des Aufsichtsrates: Dr. Hermann J. Abs. Vorstand: Dr. Herbert Culmann, Gerhard Höltje, Hans Süssenguth
Form 4064-66 (CGN KC 4) Printed in Germany

Otl Aicher: Briefbogen, ab 1962
Otl Aicher: Company letterhead, as of 1962

Otl Aicher: Flugpläne, 1967 und 1968
Otl Aicher: Timetables, 1967 and 1968

Otl Aicher: Einsteigekarten, ab 1962
Otl Aicher: Boarding passes, as of 1962

Otl Aicher: Aufkleber, ab 1966
Otl Aicher: Stickers, as of 1966

Otl Aicher: Konzeptstudie für Flugzeugbemalung Boeing 727, „Helvetica"-Schriftzug in Versalien, 1962
Otl Aicher: Concept study for Boeing 727 airplane livery, "Helvetica" logotype in all-caps, 1962

Designabteilung der Lufthansa: Leitwerk-Varianten, 1967
Lufthansa design department: Tail unit variations, 1967

Designabteilung der Lufthansa: Airbus A300 B mit „Helvetica"-Schriftzug und Bildmarke im Kreis, ab 1967
Lufthansa design department: Airbus A300 B with "Helvetica" logotype and the emblem in a circle, as of 1967

Otl Aicher/Entwicklungsgruppe 5 der HfG Ulm: Auszug aus der „Projektstudie 1400", 1962
Otl Aicher/Development Group 5 at the Ulm Design College: Extract from the "Project Study 1400," 1962

Otl Aicher: Fahrzeuggestaltung VW-Bus, 1960er Jahre
Otl Aicher: Vehicle design, VW bus, 1960s

Otl Aicher: Fahrzeuggestaltung Flugzeug-Schlepper, 1960er Jahre
Otl Aicher: Vehicle design, airplane tug, 1960s

Lufthansa

Deutsche Lufthansa Aktiengesellschaft
Flughafen-Bereich West, D-60546 Frankfurt/Main

| Ihre Zeichen<br>Your Ref. | Unsere Zeichen / Datum<br>Our Ref. / Date | Telefon / Telefax<br>Telephone / Telefax |
|---|---|---|

Gesellschaftsrechtliche Angaben,
Anschrift und weitere Informationen
auf der Rückseite
For corporate details, address and
further information p. t. o.

Designabteilung der Lufthansa nach Konzept von Zintzmeyer & Lux: Geschäftspapiere, ab 1990er Jahre
Lufthansa design department after a concept by Zintzmeyer & Lux: Design concept for Business stationery, as of 1990s

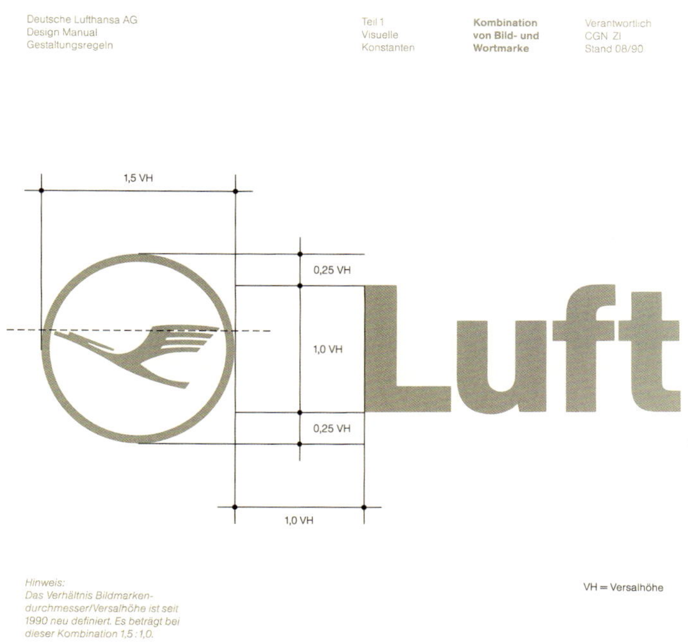


Designabteilung der Lufthansa: Design Manual, Visuelle Konstanten, 1990er Jahre
Lufthansa design department: Design Manual, visual constants, 1990s

Flugplan, 1989
Timetable, 1989

Ticket-Cover, 1990er Jahre
Ticket cover, 1990s

Zintzmeyer & Lux: Probelackierung einer Boeing 737-200 mit gelber Unterseite und gelbem Leitwerk, 1988
Zintzmeyer & Lux: Study for aircraft Boeing 737-200 livery test painting with yellow underbody and tail unit, 1988

Zintzmeyer & Lux: Modell-Studie Fahrzeug-Kennzeichnung, 1988
Zintzmeyer & Lux: Model study for vehicle markings, 1988

Überarbeitung der Bildmarke für das Leitwerk, ab 1990
Revision of the emblem on the tail unit, as of 1990

Fahrzeuggestaltung Flugzeug-Schlepper, ab 1988
Vehicle design for airplane tug, as of 1988

Fahrzeuggestaltung VW-Bus, ab 1988
Vehicle design, VW bus, as of 1988

Designabteilung der Lufthansa: Gestaltung Lufthansa Airport Express, 1982–1993
Lufthansa design department: Design for Lufthansa Airport Express, as of 1982–1993

Lufthansa
Senator
Lounge

# Die Botschafter des Fliegens

Der Komfort des Fliegens beginnt ganz offensichtlich am Boden. Von Anfang an hatte die Lufthansa Innenstadtbüros etwa in Düsseldorf, in denen Reservierungen getätigt werden konnten. Das Einchecken in den Abfertigungsterminals, an den Countern und das manchmal langwierige Aufgeben des Gepäcks, überhaupt das Warten in den Flugsteigen, durchaus auch einmal das Aushalten nennenswerter Abflugverzögerungen: All dies erfordert Räume, die den Aufenthalt und das Warten als „Transit"-Erfahrung möglichst angenehm machen. Aber auch das Warten ist und war immer gesellschaftlich hierarchisiert, wie etwa die frühen Senator Lounges der Lufthansa oder die jüngst neu eröffneten „HON Circle First Class Lounges" in Frankfurt am Main deutlich zu erkennen geben. Christian Steguweit bilanzierte 1994:

„Traditionellerweise sind die Stadtbüros aller Fluggesellschaften eher Repräsentanzen, wirken ruhig, reserviert, wenig einladend. Die Lufthansa geht hier einen neuen Weg – in konsequenter inhaltlicher Umsetzung des neuen Corporate Designs. Die Lufthansa City Center sind eine Verbindung von Stadtbüro und Reisebüro, nach aktuellen verhaltenspsychologischen Erkenntnissen gestaltet, offen, sympathisch, kundenorientiert."[1] Und an anderer Stelle betont er:

„Alle Fluggesellschaften besitzen Niederlassungen und Stadtbüros in verschiedenen Ländern der Erde. Sie sollen die Botschafter der Fluggesellschaften sein und oftmals auch das jeweilige Heimatland repräsentieren. Fluggäste sind Verbraucher, und auch sie benötigen für eine Entscheidung Stimulanz, d. h. Wünsche, Träume, Vorfreude auf Erlebnisse, wenn sie sich für ein Reiseziel oder eine Fluggesellschaft entscheiden. Von außen wirken die Stadtbüros aller Gesellschaften jedoch wenig einladend und wirken wie Repräsentanzen, nicht wie Begegnungsstätten. Sie unterscheiden sich deshalb auch nicht grundlegend voneinander, sondern sie gleichen sich. Es werden keine Träume versprochen und keine Erlebnisse verkauft. Im Gegensatz dazu präsentiert sich das Lufthansa City Center als Mischung aus Stadtbüro und Reisebüro und lädt so nach modernen psychologischen Verkaufserkenntnissen zum längeren Verweilen und Informieren ein."[2]

In den inzwischen ca. 40 Lufthansa-City-Centern – inklusive dreier Geschäfte in Italien – werden auch die entsprechenden WorldShop-Artikel angeboten. Eine Reise zu buchen und gleichzeitig Gegenstände zu kaufen, verschmilzt zu einem zusammenhängenden Erlebnis: Kauf- und Reisemotivation steigern sich gegenseitig. Inzwischen haben sich die City Center wirtschaftlich von der Lufthansa abgekoppelt. Gegenwärtig wird über Kooperationen mit der LSG und dem WorldShop nachgedacht, um das Thema „Erlebniskauf" noch weiter zu intensivieren.

Schon das erste Stadtbüro der Lufthansa, 1955 in Düsseldorf eröffnet, war sehr viel mehr als eine bloße Abfertigungsstation. Auffällige Schriftzüge und Leuchtreklamen, ein großes Bild der Lockheed Super Constellation, ein erleuchtetes Streckennetz-Diagramm an der Wand und anspruchsvolles Mobiliar unterstrichen den Anspruch des großen, lichten Raumes, ein attraktiver Anziehungspunkt für Flugwillige zu sein. Auch auf der Weltausstellung in Brüssel 1958 war das Buchungsbüro aufgrund seiner visuellen Gestaltung ein Anziehungspunkt, auch wenn dessen typografische Präsentation heute nicht nur positiv gesehen wird: „Drei leicht unterschiedliche Schriften erzeugen den Eindruck eines visuell überladenen Flugscheinbüros am Frankfurter Flughafen in den 1950er Jahren."[3]

Lufthansa City Center, ab 1991
Lufthansa City Center, as of 1991

Seite 64: Frog Design: Eingang Senator Lounge im Lufthansa Terminal auf dem Flughafen Frankfurt, 1996
Page 64: Frog Design: Entrance, Senator Lounge, Lufthansa Terminal at Frankfurt Airport, 1996

# The ambassadors of air travel

There's no question about it. The comfort and convenience of flying begin on the ground. Right from the early post-war years Lufthansa operated numerous offices – in downtown Düsseldorf, for example – where reservations could be made. Reporting to the departure terminal and checking in luggage, which can sometimes be slow, waiting on the concourse and enduring the occasional departure delay – all these pre-flight phases require areas which make waiting, as a "transit experience," as painless as possible. Waiting, too, is and has always been a function structured according to a social hierarchy clearly evidenced, for instance, in Lufthansa's earlier Senator Lounges or the newly opened "HON Circle First Class Lounges" in Frankfurt. Christian Steguweit drew up a balance in 1994:

"Traditionally, the downtown offices operated by all the airlines serve more of a representative function and are calm, reserved and not terribly inviting. Lufthansa is striking out on a new path here with consistent implementation of the new corporate design. The Lufthansa City Centers are an amalgamation of downtown reservation office and travel agency, designed in accordance with the latest findings in the behavioral sciences – open, empathetic and customer-oriented."[1] Elsewhere he emphasizes:

"All the airlines maintain local branch offices and downtown offices around the world. They are to be the carriers' ambassadors and often also represent the airline's home country. Passengers are consumers and they need stimulation in making their decision – and that means suggesting wishes, dreams and anticipation of future experiences to help them decide upon a travel destination or an airline. Seen from the

Erstes Stadtbüro in Düsseldorf, 1955
First downtown office in Düsseldorf, 1955

Flugscheinbüro auf der Weltausstellung in Brüssel, 1958
Ticketing counter at the World's Fair in Brussels, 1958

outside, the air carriers' downtown offices are not truly inviting and seem more to serve purposes of prestige than to be a place for interpersonal encounters. That is why they do not differ fundamentally one from another but indeed are all quite similar. Here no dreams are promised and no future experiences are sold. In contrast to this traditional approach, the Lufthansa City Center presents itself as a mixture of downtown reservation office and travel agency and thus makes use of modern sales psychology research to invite visitors to tarry a bit longer and obtain information."[2]

The number of Lufthansa City Centers – the revamped downtown offices – has in the meantime grown to about forty, including three in Italy. WorldShop items are also sold there. Booking a journey and at the same time purchasing products combine to create an all-round experience, while the motivations for travel and buying tend to be mutually reinforcing. In the meantime the City Centers have been spun off from Lufthansa. Thought is now being given to cooperation with LSG and WorldShop in order to further intensify the concept of shopping as a rewarding experience.

Lufthansa's first downtown office, opened in Düsseldorf in 1955, was itself far more than a simple processing point. Eye-catching lettering and illuminated advertising, a large photo of the Lockheed Super Constellation, an illuminated route map on the wall and inviting furnishings underscored the claim made by the large, light-flooded room to be an attractive drawing point for potential passengers. At the 1958 World's Fair in Brussels the reservations office was a gathering point even though, judged by today's standards, we could not give unqualified approval to the typographic concept: "Three slightly different typefaces rendered in three different ways brand a visually cramped ticket office at Frankfurt Airport Germany, 1950s."[3] Aside from the fact that the author was mistaken about the location, this interior was obviously designed prior to Otl Aicher's thoroughgoing design instructions, adopted in 1962. But this décor, even with all its inconsistencies and owing to its promoting the Super Star airliner – a further refinement of the Super Constellation, held a certain degree of charm as it exuded

Stadtbüro in San Francisco, ca. 1960/61
Downtown office in San Francisco, about 1960/61

Ticketschalter am Flughafen Frankfurt, um 1956
Ticketing counter at the Frankfurt Airport, about 1956

„Jet"-Schalter am Flughafen Frankfurt, 1960
"Jet" counter at the Frankfurt Airport, 1960

Check-in Schalter First Class am Flughafen Frankfurt, um 1970
First Class check-in counter at Frankfurt Airport, about 1970

Ticket-Counter am Flughafen München, Anfang 1990er Jahre
Ticket counter at Munich Airport, early 1990s

Frog Design: Ticket-Counter am Flughafen Frankfurt, ab 1996
Frog Design: Ticket counter at Frankfurt Airport, as of 1996

Abgesehen davon, dass der Autor sich hier im Ort irrt, war dies natürlich noch ein Auftritt vor Otl Aichers ab 1962 durchgreifenden Gestaltungsinstruktionen, aber dieser Auftritt hatte in all seinen Unvereinbarkeiten und seiner Werbung für den „Super Star", die Weiterentwicklung der Super Constellation, durchaus Charme, denn er strahlte Optimismus, Heiterkeit und Zukunftswilligkeit aus. Das Foto des Stadtbüros der Lufthansa in San Francisco lässt sich aufgrund der Wortmarke in Versalien der Schrift „Schadow Antiqua" und dem Kranich in der Parabel auf ca. 1960/61 datieren. Soweit die Innenansicht dies zeigt, war dieses Stadtbüro im vornehmen funktionalen Stil von Knoll International möbliert.

Als 1960 bei der Lufthansa das Jet-Zeitalter begann, war Schnelligkeit nicht nur in der Luft, sondern auch am Boden angesagt. Direktsprechanlagen und Rohrpostverbindungen verkürzten die Abfertigungszeit der Passagiere. Lange Reihen von Gepäckwagen in Bahnhofsuhr-Dimensionen an den Countern zeigen ebenso wie die seit Ende 1961 im Stadtbüro Frankfurt per Laufband bestätigten Platzreservierungen zum Fernschreiber das Bemühen der Fluggesellschaft, das technisch Mögliche optimal einzusetzen.

Die Flughafen-Counter der 1970er Jahre bestanden aus tiefgezogenen gelben Kunststoffboxen mit großen Radien. In den zeitgleichen Stadtbüros kamen ähnliche Verkaufs-„Quader" zum Einsatz. Dies blieb auch in den 1980er Jahren so: tiefe Volumen, auf denen die damals noch apfelsinenkistengroßen Buchungscomputer Platz fanden. Zwar wurde also die CI-Farbe Gelb sehr deutlich und flächendeckend eingesetzt und auch die Wortmarke „Lufthansa" über den Countern war dominant gestaltet, trotzdem aber überwog der Eindruck serieller Reihung. Effiziente Funktionalität war die Hauptanmutung dieser Counter. Auch die Counter der 1960er Jahre in den Flughäfen waren diesem muskulären „International Style" verpflichtet. Bis in die 1990er Jahre waren die Counter zwar oft dem generellen Design der Flughäfen angepasst, aber auf manchen Arealen, die langfristig von einer Fluggesellschaft belegt wurden, konnten auch fluglinienspezifische Gestaltungen Platz greifen, etwa in Frankfurt am Main, München oder Düsseldorf.

In den frühen 1990er Jahren kamen Bildschirme über Kopf mit Check-in- und Flugzeuginformationen hinzu, ebenso zwei Edelstahlstangen, die als Fußstütze und Gepäckablage dienten. 1995/96 folgte dann jene „postmoderne" Fragmentierung der großen Kistenformen durch gewölbte Oberflächen und fast heraldisch wirkende „Counter"-Schilder, die Hartmut Esslinger mit seinem Designbüro Frog im Rahmen eines gesamten Relaunches des Lufhansa-Auftritts entwarf. Sie wurden ab 1996 zunächst in der Abflughalle A des Frankfurter Flughafens und an den Gates B 30-33 eingesetzt. Nach 1996 gab es sie auch in Düsseldorf, Bremen, Hannover, Berlin-Tegel und im neuen Münchner Terminal 2.

Die Senator Lounges zum Überbrücken der Wartezeiten waren bis 1987 veritable großzügige Lokalitäten in Braun, Orange und Beige, bestückt mit Castiglioni-Leuchten und den vornehmen „Lounge Chairs" von Charles und Ray Eames. Andy Warhols Goethe-Siebdruck an der Wand fehlte da ebensowenig wie Rupprecht-Geiger-Siebdrucke. TV-Geräte des deutschen Unternehmens Wega, Sitzgruppen von Knoll International oder von B&B sowie schwere Marmorcouchtische vervollständigten das Ambiente, zu dem zeitgemäß ebenso 2,5 m hohe Pflanzen in Kunststoffwannen und kleine Multimirror-Birnchen in den abgehängten Decken zu gehören hatten: Lounges damals verbreiteten die schwere Gediegenheit gehobener Hotellobbys.

Nicht nur die Lounges jener Zeit, sondern auch die Flugzeugkabinen zeigten jene gedeckten Naturfarben Braun, Beige und Grün, die von Otl Aicher als „Erdfarbenkon-

optimism, cheerfulness and a commitment to the future. The photograph of Lufthansa's office in downtown San Francisco, due to the all-caps logotype in the Schadow Antiqua typeface and the crane in the parabola, can be dated at 1960/61. As far as can be seen in the interior view, this downtown office was furnished in the elegant, functional style typical of Knoll International. When the jet age started for Lufthansa in 1969, speed was of the essence not only in the air, but on the ground, too. Intercom systems and pneumatic tube conveyor systems speeded passenger processing. Long rows of luggage scales with dials the size of railroad stations clocks demonstrate the airline's efforts to make the best use of every technical innovation. In the same way, beginning at the end of 1961, operators at the Frankfurt downtown office placed ticker tape seat reservations on conveyor belts, along which they moved to the next processing point.

The airport counters in the 1970s comprised deep-drawn, yellow plastic blocks. Similar sales "cubes" were utilized in the downtown offices of the same era. This continued to be the case well into the 1980s. Deep worktops offered sufficient space for reservations computers, about the size of apple crates at that time. The yellow CI color was very apparent, being used throughout, while the Lufthansa logotype enjoyed a prominent position behind the counters. The overall impression was nonetheless of a series of individual units, all lined up in a long row. Efficient functionality was the primary impression made by these counters. Airport counters used in the 1960s were also committed to this "brawny" International Style. Until the 1990s the counters were often modified to harmonize with the particular airport's overall design. Where passenger processing areas were leased for longer periods by a single airline, however, design concepts specific to the carrier could be implemented – as was the case at Frankfurt, Munich and Düsseldorf.

In the early 1990s overhead video screens with check-in and flight information were added along with two stainless steel rails serving as a footrest or luggage rest. Following the dictates of the "post-modern," there followed in 1995/96 the fragmentation of the massive box-like shapes by incorporating curved surfaces and "counter-front shields" with their almost heraldic effect, devised by Hartmut Esslinger together with his Frog design office in the course of the overall relaunch of the Lufthansa image. They were first used in 1996 at Departure Terminal A and gates B 30-33 at Frankfurt Airport. Later they were also installed at Düsseldorf, Bremen, Hannover, Berlin Tegel and at the new Terminal 2 in Munich.

Senator Lounge am Flughafen Frankfurt, bis 1987
Senator Lounge at Frankfurt Airport, through 1987

The Senator Lounges used during stopovers and when waiting for flights were, until 1987, generously laid out facilities trimmed in brown, orange and beige, fitted with Castiglioni lamps and the elegant lounge chair by Charles and Ray Eames. Andy Warhol's "Goethe" silkscreen was as omnipresent as Ruprecht Geiger prints. Television sets made by the German manufacturer Wega, seating groups by Knoll International or B&B as well as the weighty marble coffee tables rounded out the atmosphere. Also on hand, reflecting contemporary tastes, were houseplants 2.5 meters tall, rooted in plastic tubs, and small, multi-reflector halogen spotlights in the suspended ceilings. Lounges at that time suggested the classical elegance of luxury hotel lobbies.

Not only the lounges of that era, but the passenger cabins, as well, featured those restrained natural colors of brown, beige and green which Otl Aicher introduced into the Lufthansa corporate design as the "earth tones concept." The mental ambivalence inherent to flying was to be counteracted with a feeling of being firmly rooted on earth.

All this changed significantly at the end of the 1980s. The Zintzmeyer & Lux agency, located in Zurich, was charged with overhauling the airline's corporate identity. They made the Senator Lounges far cooler, with elegant gray panels, used silver trim against

Zintzmeyer & Lux: Senator Lounge, bis 1996
Zintzmeyer & Lux: Senator Lounge, through 1996

Zintzmeyer & Lux: Frequent Traveller Lounge, bis 1996
Zintzmeyer & Lux: Frequent Traveller Lounge, through 1996

zept" ins Corporate Design der Lufthansa eingeführt wurden. Die seelische Ambivalenz des Fliegens sollte durch die Anmutung von Erdverbundenheit ponderiert werden.

Am Ende der 1980er Jahre änderte sich dies deutlich. Die damals mit der Überarbeitung der generellen Corporate Identity der Fluggesellschaft beauftragte Zürcher Agentur Zintzmeyer & Lux gestaltete nun die Senator Lounges wesentlich kühler, mit vornehmen grauen Paneelen, mit silberfarbenen Dekoren auf Grau bzw. in der Vielflieger-Lounge mit blauem Dekor auf Grau und dezenten Farbhinweisen auf die Corporate-Identity-Farbe Melonengelb. Die Sitzreihen kamen von Wilkhahn, die freistehenden Stühle von Vitra. Freistehende Sessel gab es nicht. Und weltweit zierten jede LH-Lounge ein Flugzeug-Großfoto und eine persönliche Textansprache der Kunden. Der „Hotel-Touch" wurde also eher durch einen „Messe-Touch" ersetzt: kühler und rationaler, eben businesslike! Dieses Konzept galt bis 1996. Die Abflugbereiche der Lufthansa dieser Zeit zeigen ein ähnliches, allerdings weniger einer Clubatmosphäre zugeneigtes Ambiente: mit Sesseln der Firma Vitra in linearer Reihung mit durchaus hohem Komfort. Auch die Counter dieser Phase vermittelten den Eindruck kühler Messearchitektur: verlässlich, sicher, kompetent.

1994 heißt es in einer Standortbestimmung der damaligen Corporate Identity:

„[Die] Lufthansa arbeitet außerdem an einem Betreuungskonzept ‚am Boden' (vor dem Flug), das den unterschiedlichen Bedürfnissen der Geschäftsreisenden mehr entspricht und stärker als in der Vergangenheit individuelle Betreuung auch in den Lounges zuläßt. Da je nach Reiseanlaß heute mehr als früher unterschiedliche Reiseklassen genutzt werden, besteht das neue System aus einer sehr komplexen Betreuungsmatrix. Ziel ist es, den Lufthansa-Kunden lästige Wartezeiten und kompliziertes Handling zu ersparen. Bei Geschäftsreisenden und Vielfliegern würde sich die Lufthansa damit zum einen durch eine spezifizierte Betreuungsqualität, zum anderen aber auch durch einen unverwechselbaren formalen Rahmen, der Lounges von Business- bis Senator-Class-Reisenden enthält, von anderen Gesellschaften differenzieren. Unternehmensidentität und Betreuungsleistung verschmelzen zu einem gemeinsamen Ganzen. Lufthansa bemüht sich deshalb auch im Ausland, schon etwas wie ‚vorweggenommene' Heimat zu symbolisieren …"[4]

Kurze Zeit später wehte der Zeitgeist erneut anders: Jetzt wollte man die inzwischen etwas konservativ wirkende Gediegenheit verlassen und erneut auf den alten Pioniergeist setzen. Dynamik, Abenteuerlust, Emotion waren um 1995 die Begriffe der Stunde. Ein Designer dafür war schnell gefunden: der agile Hartmut Esslinger mit seinem kalifornischen Designstudio Frog. Nun ging es entsprechend einem oft zitierten Motto von Esslinger nicht mehr um jene „form follows function"-Auffassung der Otl-Aicher-Phase, sondern um „form follows emotion". „High-Touch" also statt „High-Tech" hieß nun die Devise. An diesem Auftrag der Image-Renovierung oder Repositionierung der emotionalen Werte der Corporate Identity arbeiteten die Teams von Frog in Altensteig im Schwarzwald sowie in Sunnyvale/Kalifornien. Als roter Faden des Relaunches der Anmutungsqualitäten der Fluglinie wählten die Frog-Designer das beplankte Wellblech des legendären Vorkriegsflugzeuges Ju 52, das erstmals 1931 in Dessau gebaut wurde. Unter dem Titel „Abenteuerlicher Auftritt" bilanzierte die Fachzeitschrift *Design Report* Mitte 1997:

„Leicht geschwungene Wellen schlängeln sich durch das gesamte Programm: mattes, kurvenförmiges Aluminium-Wellblech teilt in Hüfthöhe die mit Buffet, Getränkeautomat [und Bistro-Möblierung] ausgestattete Business-Wartezone vom Economy-Bereich, dessen einziger Luxus in den grauledernen „ARC"-Bänken von König & Neurath besteht. Moderne Formen vollziehen hier – schwungvoll zwar, doch unzeit-

gray or in the Frequent Traveler Lounge, blue trim against gray, accented with restrained hints of the melon yellow from the corporate livery. Row seating was made by Wilkhahn while the free-standing chairs were Vitra models. There were no freestanding armchairs. And all around the world every Lufthansa lounge was graced with a large aircraft photo and a personal message for passengers. The "hotel touch" thus gave way to something more like a "trade fair look" that was cooler and more "rational" – and thoroughly business-like! This concept was used through 1996. The departure areas at the Lufthansa gates in the mid-nineties were similar in their ambiance, but with a rather less "clubby" atmosphere. Bench seats made by Vitra and arranged in rows offered a high degree of comfort. The counters during this phase also projected a feeling reminiscent of cool trade fair architecture: reliable, sure, competent.

In 1994, in an analysis of the corporate identity at that time, it was found that:

"Lufthansa is also working on a customer care concept 'on the ground' (before the flight) that better corresponds to the varying requirements of business travelers and, more than in the past, permits individual attention to passenger needs in the lounges, as well. In contrast to past conventions, today's passengers ever more frequently book a particular class of service in accordance with the reason for traveling. As a consequence, the new system features a complex customer care matrix. The goal is to save Lufthansa passengers annoying waiting periods and complex processing procedures. Thus Lufthansa intends to set itself apart from other carriers in the eyes of business travelers and frequent fliers. This is to be achieved with a specific quality in customer care, on the one hand, but also with an unmistakable formal framework, which includes the lounges for Business and Senator Class travelers. Corporate identity and customer care merge to create a unified whole. It is for this reason that Lufthansa attempts, in foreign fields as well, to symbolize something like 'a bit of home away from home' …" [4]

A short time later the *zeitgeist* had taken yet another turn. The aim now was to depart from the elegant look which in the meantime had come to be seen as rather conservative, and to return once more to the old pioneering spirit. Dynamics, a sense of adventure and highly charged emotions were the concepts of the hour around 1995. The right designer was quickly found: the agile Hartmut Esslinger with his Frog design studio in California. Now, following an often-quoted Esslinger credo, it was no longer a question of "form follows function" – the dominant approach during the Otl Aicher phase – but rather of "form follows emotion." The goal now was "high touch" instead of high-tech. The Frog teams at Altensteig in the Black Forest and in Sunnyvale, California, worked on this commission to revamp the image and reposition the emotional values embodied in the airline's corporate identity. When selecting a visual constant to be used in the relaunch of the airline's general image the Frog designers selected the corrugated sheet metal panels found in the legendary JU 52 airframe, first built in Dessau in 1931. Under the title "Adventurous stage effects" the *Design Report* journal drew up a balance at mid-1997:

Frog Design: Empfangscounter Business Lounge am Flughafen Frankfurt, ab 1996
Frog Design: Business Lounge reception desk at Frankfurt Airport, as of 1996

"Gently undulating waves wend their way through the entire program. Mat, waist-high, and curved corrugated aluminum dividers separate the business waiting area – equipped with a buffet, beverage dispenser [and bistro-type furnishings] – from the economy area, whose only luxury is found in the gray leather upholstered 'ARC' benches by König & Neurath. Modern forms – indeed dynamic but nonetheless out of phase with the times – denote class divisions. At the new counters, by contrast, the corrugated surfaces and the curved front panels bring the passengers and employees closer together and permit a direct view into the check-in procedures. Between the overhead screens, too, corrugated sheet steel decoration in the shape of a horizontal

Frog Design: Senator Lounge, ab 1996
Frog Design: Senator Lounge, as of 1996

gemäß – eine Klassentrennung. An den neuen Schaltern dagegen führen die gewellten Oberflächen und gebogenen Fronten Passagiere und Bedienstete näher zusammen und erlauben einen direkten Einblick in die Abläufe. Auch zwischen den Überkopf-Bildschirmen faßt eine gewellte Stahlblechapplikation in Form eines horizontalen Bandes die Geräte optisch zusammen. Und selbst das neue Leitsystem schlängelt sich von der Decke herab dem Passagier entgegen."[5]

Die von Frog gestalteten Lounges sahen etwa in der Business Class eine Bar und eine Videowand vor, Konferenz- und Ruhezonen, aber auch Workstations mit Faxanschluss. Die Lounge der ersten Klasse zeigt ebenso unterschiedliche Arbeits- und Ruhebereiche sowie eine Erfrischungsbar. Postmoderne Segmentbogen-Trennscheiben konturieren Räume. Die Eingänge zu den Lounges, durchaus nicht zurückhaltend wellblechverplankt, zieren ovale Fenster in Flügeltüren, die den Bullaugen der Ju 52 nachempfunden sind. Bild- und Wortmarke sind nun konsequent gelb, auf Blau wird hier verzichtet. Man sitzt in der Business Class auf grauledernen Clubsesseln oder auf grauen Arne-Jacobsen-Stühlen, in der Senator Lounge auf blauen Sesseln. Die geschwungenen Trennwände ermöglichen abgeschirmte digitale Kommunikation. Aber die mit Kappnähten konturierten Clubsessel und die grazilen Einbeiner-Glastische in der Senator Lounge erinnern an TV-Sendungen, in denen Moderator und Gast in ähnlichen Zweierarrangements agieren. Anmutungspiraterie: Die Formen, Farben und Anordnungen suggerieren einen per se öffentlichkeitswirksamen Auftritt! Geradezu meditativ geriet dann den Frog-Designern aber die „Time Tunnel" genannte Verbindung von Flugsteig A zu B am Frankfurter Flughafen. Die Frösche verpassten diesem zum Lufthansa-Bereich gehörenden Tunnel wechselndes magisches Licht, changierend von Violett zu Grün, unterschiedliche Musikberieselung und ebenso unterschiedliche Ge-

Frog Design: Tunnel zwischen Gate A und B am Flughafen Frankfurt, 1996
Frog Design: Tunnel between Gates A and B at Frankfurt Airport, 1996

rüche. So werden etwa Brasilien-Reisende bereits hier mit Samba-Rhythmen und tropischen Farben auf ihr südamerikanisches Erlebnis eingestimmt.

Aber diese „Emotionsoffensive" war offenbar nach der Jahrtausendwende nicht mehr so ohne weiteres zeitgeistkompatibel. Das ab 2001 beauftragte Designbüro Bernd Hollin und Alexander Radoske entwickelte ein im Vergleich zu den Frog-Designern differentes Attraktionsraster sowohl für First Class Terminals und Lounges als auch für Senator Lounges. Zunächst wurden neue Senator Lounges in München realisiert, dann folgte ab 2004 die HON Circle First Class Lounge in Frankfurt am Main. Hier dominiert ein minimalistisches, rektangulär-nüchternes Raum- und Ausstattungskonzept in war-

strip draws these devices together visually. And even the new passenger guidance system snakes down from the ceiling, toward the traveler." [5]

The lounges designed by Frog provided special amenities for Business Class passengers: a bar and a video wall, conference areas and quiet zones, and workstations featuring fax service. The First Class lounge also provided distinct working and relaxation areas and a refreshment bar. Post-modern, arc-shaped divider panels lent contour to the spaces. The entrances to the lounges, by no means restrained in their use of the corrugated metal panels, featured oval windows in swinging doors, reminiscent of the "portholes" of the JU 52. The emblem and logotype here were in yellow throughout; the design does without any blue at all. Business Class passengers had a choice of gray leather club chairs or gray Arne Jacobsen chairs while blue armchairs were provided in the Senator Lounge. The curved dividing panels provided privacy in digital communications. The armchairs contoured with double rows of stitching and the graceful glass-topped pedestal tables in the Senator Lounge reminded observers of television programs with the moderator and guests seated in similarly paired arrangements. This was clearly an "act of piracy": the shapes, colors and arrangements suggest in themselves a setting directed toward a viewing audience! By contrast, the Frog designers cast the so-called "Time Tunnel" – the connection from Concourse A to Concourse B at the Frankfurt Airport – in thoroughly meditative tones. They fitted this passageway, which is within Lufthansa's domain, with an almost magical light, changing from violet to green, complemented with a selection of background music and varied scents. In this way, for example, travelers to Brazil are put in the mood for their South American experience with samba rhythms and tropical colors.

But this "emotion-based campaign" was apparently not entirely compatible with the spirit of the new millennium. The design office of Bernd Hollin and Alexander Radoske was put under contract as of 2001 and their designers developed, when compared with the Frog designers, a differentiated matrix of attractions for First Class terminals and

Bernd Hollin, Alexander Radoske: Senator Lounge Flughafen München, 2002
Bernd Hollin, Alexander Radoske: Senator Lounge at Munich Airport, 2002

men Farben, das durch strategisch klug eingesetzte Lichtbänder dynamisiert wird. Viel Eichenholz, patinierte Messingbleche, Teppiche, große Glasflächen als Abgrenzungen, skulpturale rektanguläre braune Ledersitze, ein „High-Touch"-Ambiente in minimalistischer Gestaltreduktion. Raum- und Einrichtungskonzepte für eine metropolitane Klientel, der jede ornamental-emotionale Ablenkung wie auch jeder modische Schnickschnack zu viel wäre. Das prinzipielle Designkonzept von Hollin und Radoske für das spezielle Terminal und die HON Circle First Class Lounge entstand in enger Zusammenarbeit mit der Frankfurter Dependance der Hamburger Agentur Peter Schmidt Group. Hier zeigt sich eine aktuelle Designsprache, die ihre Mittel, Materialien und Technologien mit einem souveränen Selbstbewusstsein einsetzt, das Askese mit Luxus zu amalgamieren weiß. Und genau dies steht der Produktsprache der Lufthansa in der unmittelbaren Gegenwart gut an, denn eine solche Produktsprache vermittelt, dass diese Fluggesellschaft zwar rational, solide und funktional ist, gleichzeitig aber von einer metropolitanen Eleganz zehren kann. Ein solches Ambiente unterfüttert gewissermaßen die Rationalität emotional.

Das Design der HON Circle First Class Lounges ist modular aufgebaut und kann an unterschiedlichste Raumsituationen angepasst werden. Die Module bestehen aus einem Eingangs- sowie einem Empfangsbereich inklusive Gepäckservice und Garderobe; einem diskreten Service Center mit LH-Mitarbeitern für Buchungsfragen; aus einer Komfortzone mit Sitzgruppen und hochwertigen Teppichen, Marmorraumteilern und transluzenten hängenden Wänden; aus einem beruhigten Relaxbereich mit Relaxsesseln mit integrierten Medienboxen; aus einem Sanitärbereich mit Dusch- und Wannenbädern sowie einem separierten Bereich für eine Hausdame; einer großzügigen Regenerationsküche, in der warme Speisen auf Verzehrtemperatur gehalten oder vom gekühlten Zustand auf entsprechende Temperaturen gebracht werden; einer Cigar-Lounge, dem Raucherbereich der Komfortzone; einem Barbereich als kommunikativem

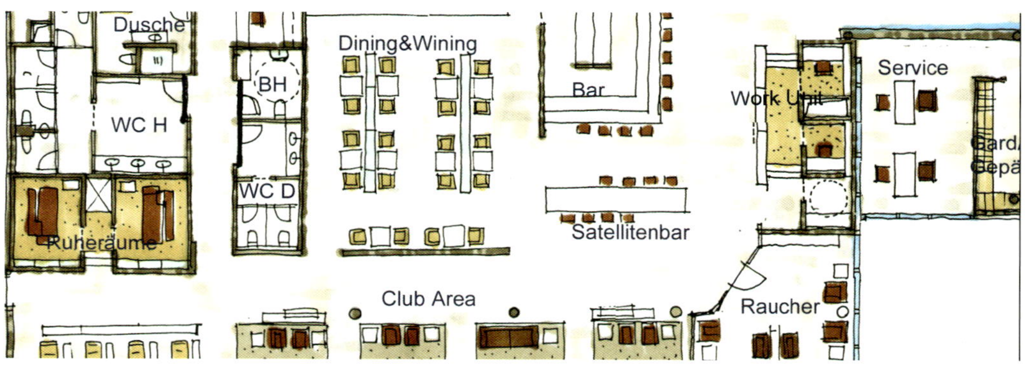


Bernd Hollin, Alexander Radoske: HON Circle First Class Lounge, Grundrißstudie, 2004
Bernd Hollin, Alexander Radoske: HON Circle First Class Lounge, draft floor plan, 2004

Zentrum in der Nähe des Lounge-Eingangs; aus Office Units mit mehreren räumlich getrennten Arbeitsplätzen; einem Restaurant gehobener Klasse in Ausstattung und Service mit einem SB-Buffet, genannt „Dining + Wining" sowie einem Ruheraum mit abgetrennten und abschließbaren Liegebereichen.

Das seit 1. Dezember 2004 geöffnete First Class Terminal ähnelt im Ambiente eher einem Nobelhotel als einer schnöden Abflughalle. Die Kontrolle von Tickets und Pässen erfolgt in diesem Terminal. Ein Gourmetrestaurant, eine edle Bar, großzügige Badeinheiten sowie Arbeits- und Ruheräume sind angegliedert. Der neue „Top-Service für Top-

Lounges and for the Senator Lounges. The redesigned Senator Lounges were first installed at Munich and these were followed in 2004 by the HON Circle First Class Lounge at Frankfurt. Dominant here is a minimalist, soberly rectangular concept for spaces and equipment, using warm colors. Illuminated strips cleverly placed at strategic locations lend a more dynamic feeling. Oak trim, antiqued brass and carpeting are lavishly utilized, large glass panels are used as dividers, and sculpted, rectangular, brown leather seats accentuate a high-touch feeling in a minimalist design reduction. This is a concept for space and furnishings tailored to a metropolitan clientele, for which every emotional and ornamental distraction and every trendy gimmick would be too much. The fundamental design concept for the special terminal and HON Circle First Class Lounge was devised by Hollin and Radoske in close cooperation with the Frankfurt office of the Hamburg-based Peter Schmidt Group. Seen here is a contemporary design vernacular which employs its resources, materials and technologies with a supreme self-awareness and which is successful in blending asceticism with luxury. And this is exactly in tune with the product language at Lufthansa, firmly anchored in today's world, a language suggesting that this carrier is, indeed, efficient, reliable and highly functional but one which can at the same time draw upon metropolitan elegance. A setting such as this lends emotional reinforcement to the spirit of rationalism.

The design for the HON Circle First Class Lounge is modular in its structure and can be modified to suit a wide variety of spaces. The modules comprise an entryway and reception area – including baggage service and a cloakroom; a separate service center for discussing flights with Lufthansa staff; a "retreat" zone with seating groups and high-quality carpets; marble room dividers and translucent suspended panels; a subdued relaxation area with thickly upholstered lounge chairs featuring integrated audio-visual servers; bath facilities with both showers and bathtubs and a separate area for the attendant; a generous warming kitchen; a cigar lounge – the designated smoking area in the retreat zone; a bar as the communicative center near the lounge entrance; office units with several workstations isolated one from another; a restaurant with superior appointments and service and also providing a self-service buffet – known as "Dining + Wining," and a sleeping area featuring separate and lockable cubicles for naps.

In terms of its overall feeling, the First Class Terminal at Frankfurt, opened on December 1, 2004, more resembles a luxury hotel than a plain and simple departure hall. Tickets and passports are checked inside this terminal building. A gourmet restaurant, an elegant bar, generously dimensioned bathing facilities and working and relaxation areas are included. The new "top service for top clients," occupying 1,800 square meters of floor space, also offers a Porsche Cayenne or a Mercedes S-Class sedan with which the well-heeled travelers are whisked to the aircraft. The passenger's needs are attended to throughout by a personal assistant who helps during check-in, at security inspections and with customs formalities. On average, 350 passengers a day are to be pampered in the new terminal. A further First Class terminal is scheduled to open in Munich at the beginning of 2006.

Lufthansa's response to ever tougher competition in the skies is product differentiation. Offering the entire range – from economy, business and first class tickets and on through to luxury travel – appears to Lufthansa to be the most promising way to remain competitive at present.

In addition to newly designed counters and lounges, automated check-in procedures are also to be found among new amenities introduced on the ground over the course of recent years. The first such machines were installed in April of 1994, initially at the Cologne/Bonn, Düsseldorf and Hanover airports. This new service speeds processing

Kunden“ auf 1.800 Quadratmetern beinhaltet auch die Bereitstellung eines Porsche Cayenne oder einer Mercedes-S-Klasse-Limousine, mit der die Reisenden zum Flugzeug gefahren werden. Betreut wird der Fluggast währenddessen von einem persönlichen Assistenten, der beim Check-in, bei den Sicherheitskontrollen und bei Zollformalitäten behilflich ist. Im Schnitt sollen 350 Fluggäste täglich im neuen Terminal verwöhnt werden. Für Anfang 2006 ist ein weiteres First Class Terminal in München geplant.

Im härter werdenden Wettbewerb am Himmel geht es auch bei der Lufthansa um Produktdifferenzierung. Das gesamte Sortiment von Economy-, Business- und First-Class-Tickets bis zu Luxuspassagen anzubieten, erscheint der Lufthansa gegenwärtig die vielversprechendste Möglichkeit zu sein, konkurrenzfähig zu bleiben.

Aber neben Countern und Lounges gehören zu den im Laufe der Jahre neu hinzukommenden Annehmlichkeiten für die Fluggäste am Boden z. B. auch Check-in-Automaten, deren erste im April 1994 zunächst auf den Flughäfen in Köln/Bonn, Düsseldorf und Hannover in Betrieb genommen wurden. Dieser neue Service verkürzte die Abfertigung an Flughäfen erheblich. Mit Kreditkartenbezahlung kann sich der Passagier seinen Flugschein ausstellen lassen und gleichzeitig einchecken: So entfallen Wartezeiten am Schalter. Nach und nach stellte die Fluglinie ihre herkömmlichen Flugscheine vollständig auf diese „ATB“-Tickets (Automated Ticket and Boarding Pass) genannten Flugscheine um. Bei ihnen sind alle Daten auf einem Magnetstreifen auf der Rückseite gespeichert. Der Dialog mit dem Terminal erfolgte früher durch Tasten, heute durch Berühren des Bildschirms, also im sogenannten „Touch Screen“-Verfahren. Dieses Check-in-System entwickelte das Offenbacher Designbüro „Vision & Gestalt“ (Bernd Bürdek und Stephan Schupach). Dabei entwickelte das Entwerferteam diese Automaten in Zusammenhang mit weiterreichenden Forschungen über das sogenannte „Human Interface Design“, das sich u. a. zur Aufgabe gestellt hat, die Unanschaulichkeit der Technik, die sich oftmals in kryptischen Bedienungsanweisungen äußert, nachvollziehbarer und einsehbarer zu machen. In der ersten Hälfte der 1990er Jahre setzte sich die grafische Gestaltung von Computeroberflächen durch und breitete sich immer weiter aus. Dabei muss dieses Dialogverfahren, mit dem der Benutzer am Bildschirm operiert, ebenso eine Selbsterklärungsfähigkeit besitzen wie eine Erwartungskonformität, zu schweigen von Fehlertoleranz und Fehlertransparenz. Zwischenzeitlich haben sich diese Selbstbedienungsterminals bewährt: nicht nur aufgrund ihrer nachvollziehbar logischen Software-Benutzeroberfläche, sondern sicherlich auch deshalb, weil inzwischen die Flugpassagiere wohl mehrheitlich langgewachsene Erfahrungen im Umgang mit Computersoftware haben. Und natürlich werden auch diese Check-in-Automaten in ihrer Farb- und Grafikgestaltung dem Erscheinungsbild der Lufthansa angepasst, ebenso wie die Boarding-Automaten.

Es sollte deutlich geworden sein, dass ein Passagierflug sehr viel mehr ist als die Passage von A nach B. Buchen, einchecken, Aufenthalt und Zwischenaufenthalt müssen komfortabel und angenehm organisiert und gestaltet werden. Der Service in der Luft und der Service am Boden sind dialektisch aufeinander bezogen: Erst ihre Einheit konfiguriert das Flugerlebnis.

**1** Christian Steguweit, „Lufthansa. Typologie und Konsequenz einer Corporate Identity“, in: Klaus Schmidt (Hrsg.), *Corporate Identity in Europa: Strategien, Instrumente, Erfolgreiche Beispiele*, Frankfurt am Main/New York 1994, S. 135–142, S. 141.
**2** Christian Steguweit, in: Norbert Hammer, *Die stillen Designer – Manager des Designs*, Essen 1994, S. 110 f.
**3** Keith Lovegrove, *Airline. Identity, Design and Culture*, New York 2000, S. 125.
**4** Christian Steguweit, „Corporate Identity – Problem oder Lösung für die Fragen der 90er Jahre – Eine Standortbestimmung am Beispiel der Lufthansa“, in: Norbert Hammer, a. a. O., S. 103–116, S. 112.
**5** Birgit Müller, „Abenteuerlicher Auftritt“, *Design Report*, Nr. 5, 1997, S. 39.

at the airport considerably. Paying by credit card, the passenger can have the ticket issued and check in at the same time, eliminating waiting at the counter. The airline has gradually made the complete transition from conventional tickets to this "ATB" (automated ticket and boarding pass) system. Here all the requisite data are recorded on a magnetic strip on the back of the pass. Whereas a keyboard was initially used as the input device at the computer terminal, touchscreens are now employed for the dialog-driven inter-face. This check-in system was engineered by the Vision & Gestalt design office in Offenbach (Bernd Bürdek and Stephan Schupach). The development team took account of advanced research in human interface design when planning these terminals. The objective here is to replace existing technology – sometimes virtually impenetrable and often explained with highly cryptic instructions – with techniques which are more easily grasped. In the first half of the 1990s the graphic user interface made its breakthrough and is becoming ever more prevalent. This dialog-based process, in which the user works at a screen, not only has to be self-explanatory; it must also conform to user expectations and be highly error-tolerant while offering full transparency in regard to error corrections. These self-service terminals have proven their usefulness in the meantime. This is certainly due not only to their logical and easily understood software interface but most assuredly because the majority of travelers have by now amassed extensive experience in dealing with computerized systems. And it goes almost without saying that the colors and graphic designs at these automatic check-in terminals are coordinated with the Lufthansa corporate image. This is also true for the automatic processing machines at the boarding door.

All this should have made it quite clear that the passenger sees flying as far more than just getting from point A to point B. Making reservations, checking in, and periods of waiting prior to departure and at intermediate stops have to be organized and designed so as to be as comfortable and pleasant as possible. Service in the air and on the ground are dialectically referenced one to another. Only when they achieve unity is the flying experience properly configured.

Check-in-Automat Flughafen Frankfurt, 2000 bis heute
Automated check-in terminal at Frankfurt Airport, 2000 to the present

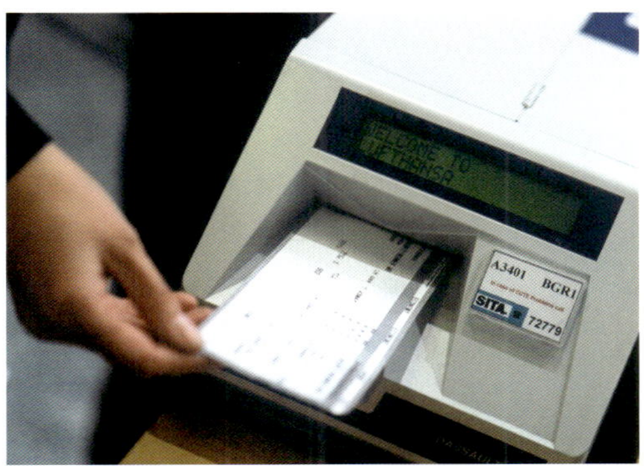


Boarding-Automat, Mitte 1990er Jahre bis heute
Automated check-in terminal, mid-1990s to the present

**1** Christian Steguweit, "Lufthansa. Typologie und Konsequenz einer Corporate Identity," in: Klaus Schmidt (ed.), *Corporate Identity in Europe: Strategien, Instrumente, Erfolgreiche Beispiele*, Frankfurt/New York, 1994, pp. 135 to 142, here p. 141.
**2** Christian Steguweit, in: Norbert Hammer, *Die stillen Designer – Manager des Designs*, Essen, 1994, pp. 110 f.
**3** Keith Lovegrove, *Airline. Identity, Design and Culture*, New York 2000, p. 125.
**4** Christian Steguweit, "Corporate Identity – Problem oder Lösung für die Fragen der 90er Jahre – Eine Standortbestimmung am Beispiel der Lufthansa," in: Norbert Hammer, op. cit., pp. 103–116, here p. 112.
**5** Birgit Müller, "Abenteuerlicher Auftritt," *Design Report*, no. 5, 1997, p. 39.

Bernd Hollin, Alexander Radoske: HON Circle First Class Lounge, Flughafen Frankfurt 2004
Bernd Hollin, Alexander Radoske: HON Circle First Class Lounge, Frankfurt Airport, 2004

Bernd Hollin, Alexander Radoske: First Class Terminal, Flughafen Frankfurt, 2004; Empfang, Komfort Zone, Relax Bereich, Dining & Wining, Badinterieur
Bernd Hollin, Alexander Radoske: First Class Terminal, Frankfurt Airport, 2004; reception, comfort zone, relaxation area, Dining & Wining, bath interiors

DO NOT ENTER

# Relaxen über den Wolken

Die wichtigsten Kriterien der „Flugtauglichkeit" von Bordgeschirren, Fluggastsitzen, Wand- und Bodenverkleidungen sowie Nasszellen sind Gewicht und Brandschutz. In keinem anderen Großraum-Verkehrsmittel, ob Schiff oder Bahn, sind die entsprechenden Anforderungen und Vorschriften so rigide und müssen es sein. So ist etwa das wabenartige High-Tech-Material „Honey Comb" äußerst leicht, äußerst stabil, aber auch äußerst teuer. Trotzdem wird aus diesem Material die Innenraumverkleidung vieler Flugzeuge gefertigt.

Zudem erfordert die Innenraumgestaltung der Flugzeuge eine geradezu ingenieurwissenschaftliche Präzision und eine extreme Passgenauigkeit aller Einzelteile. Trotzdem oder gerade deshalb werden Formen, Farben, Grafik, Texturen und Stoffe tendenziell eher dazu genutzt, den kühlen High-Tech-Charakter der Flugzeuge – ähnlich wie bei Automobilausstattungen – zu konterkarieren. Dabei sind der Simulation privater Wohnumwelten natürlich Grenzen gesetzt: zunächst durch den langgestreckten, zigarrenförmigen Innenraum, selbst wenn er zweigeschossig ist wie bei den großen Interkontinentalflugzeugen; dann durch die Serialität der Sitz- und Fensterreihen; schließlich durch begrenzte Raumkapazität und die sehr strengen Anforderungen an die Sicherheit. Natürlich spiegeln die Interieurs trotzdem den jeweiligen Zeitgeist und die „Welt da draußen". Vieles ist zudem von den Herstellern der Flugzeuge vorgegeben: die Form der Fenster und der Wandverkleidungen, fast immer die Anordnung der Sitze, die nur in Grenzen variabel ist, die Form der Gepäckfächer über den Sitzen, die Position der Nassräume in der Maschine, obwohl gerade diese in den heutigen Flotten immer mehr frei positionierbar werden. Die Implantierung der fluggesellschaftsspezifischen Corporate Identity konzentriert sich infolge dessen mehr auf die Oberflächen- und Farbgestaltung des Passagierraumes, auf grafische Schrift- und Bildelemente auf den Stirnseiten am Ende der verschiedenen Fluggastabteile, auf die Stoff- und Texturauswahl der Sitze und Bodenbeläge. Am ehesten noch werden die Gestaltung der Fluggastsitze und jene der Nassräume an externe Designer und Designbüros vergeben.

Die Ausstattung der Lufthansa-Maschinen war in ihren Anfangsjahren noch ebenso bieder wie jene der zeitgleichen privaten Haushalte. Trennwände zeigten altdeutsche Merian-Stadtansichten und natürlich den Lufthansa-Kranich, Servietten hatten auch schon einmal Küchenhandtuchkaros und man saß sich an tatsächlichen Tischen traut gegenüber wie in manchen Flugzeugen der 1930er oder 1940er Jahre. Raumknappheit wie in späteren Jahrzehnten mit ihrer massentauglichen Komprimierung der Sitzreihen war noch nicht notwendig, denn noch war Fliegen ein exklusives Privileg. Auch in den 1960er Jahren zeigen die Ausstattungen noch viel freie Raumteile und man konnte sich noch eine weit größere Beinfreiheit in allen Flugklassen vor den Sitzen erlauben als heute. Trotzdem bleibt bis heute wahr: Die dominante Tätigkeit des Passagiers ist das Sitzen – je nach Länge des Fluges zwischen einer und über zehn Stunden. Sobald man sich an anderen Passagieren vorbeigedrückt hat, sinkt man aufatmend in seinen Sitz, schließt den Sicherheitsgurt, studiert die Safety-Karten, lauscht aufmerksam den Notfallerläuterungen der gestikulierenden Flugbegleiter, blättert im Bord-Magazin aus der Rückentasche des Vordersitzes, späht aufs Rollfeld und wartet auf den Start. Sobald dieser eingeleitet ist, wird man mit 280 bzw. 350 Stundenkilometern in seinen Sitz gedrückt und genießt zum ersten Mal den Komfort dieses Sitzmöbels, dem man sich fürderhin für Stunden anvertraut.

Seite 82: Treppe zur Underfloor-Galley des Lufthansa Airbus A340-600, ab 2004

Page 82: Stairs to the underfloor galley at the Lufthansa Airbus A340-600, as of 2004

# Relaxing above the clouds

Weight and fire safety are unquestionably the most important criteria for the "airworthiness" of airline dinnerware, passenger seats, wall and floor coverings, and service areas. For no other large-scale vehicles, whether on rails or on the water, are safety requirements and regulations so stringent, but they are a simple matter of necessity in aviation. One example of a response is the high-tech "honeycomb" material. It is extremely light, extremely stable, but also extremely expensive. In spite of this, the interior claddings for many airplanes are made from this material.

Moreover, designing aircraft interiors demands all the precision of the engineering sciences and extremely close fits for all the individual elements. And in spite of – or exactly because of – this the shapes, colors, graphics, textures and materials tend to be employed more to offset the cool, technological character of the airplane. A similar the situation is found in motor vehicle interiors. Yet simulating private residential spaces is subject to certain limits. The first such limitation is the long, tube-like interior, even where it is divided into two levels as in large intercontinental aircraft. Then comes the regularity of the rows of seats and windows. The final considerations are the limited space available and the extremely rigid safety requirements. Nonetheless, the interiors reflect the current *zeitgeist* and the feeling of the "wide, wide world." A great deal is determined by the aircraft manufacturers: the shape of the windows and wall panels, the arrangement of the seats, which can be varied only within certain limits, the shape of the luggage bins above the seats, and the location of the toilets within the aircraft – even though they can be more freely positioned in today's fleets. Efforts to impose the corporate identity specific to the airline thus concentrate more on the design of surfaces and colors in the passenger compartment, on graphically interesting copy and images on the bulkheads separating the various passenger compartments, and on the selection of fabrics and textures for the seats and floor coverings. Designing the passenger seats and service areas are the assignments most likely to be contracted to outside designers and design offices.

The appointments in Lufthansa aircraft in its early years were just as pedestrian as those in private households at the time. Partitions were decorated with medieval depiction of cities and, of course, the Lufthansa crane. Dinner napkins sometimes had a checkered kitchen-towel pattern and passengers actually sat face-to-face across real tables in some aircraft in the 1930s or 1940s. Close quarters such as those found in later decades with their "compression" of the seat rows in order to handle mass travel was not yet necessary. Flying was still a very exclusive privilege at that time. Even in the 1960s the layouts still included a large amount of free space and carriers could afford far more legroom in every class than is the case today. Nevertheless, the basic situation remains unchanged: the passenger's primary activity is sitting – for between one and more than ten hours, depending on the distance covered. Having squeezed by others who are boarding, the passenger settles into the seat with a sigh of relief, buckles the safety belt, studies the safety brochure, listens carefully as gesticulating flight attendants explain the emergency procedures, browses through the inflight magazine extracted from the pocket on the seat in front, inspects the runway through the windows, and waits for departure. Once take-off is started the passenger is pressed into the seat at 280 or 350 kilometers per hour and experiences for the first time the comfort of this "furniture" to which he has entrusted himself and his well-being over the next few hours.

Kabine einer Lockheed Super Constellation der Lufthansa, 1955–1967
Passenger cabin in a Lockheed Super Constellation operated by Lufthansa, 1955–1967

Auch Christian Steguweit, bis 1994 Leiter Brand Management und Corporate Identity bei der Deutschen Lufthansa, betont:

„Die Reise, insbesondere bei Langstreckenflügen, reduziert sich auf eine ganz wesentliche Funktion, auf die Bequemlichkeit zu sitzen. In der Werbung versprechen fast alle Fluggesellschaften Beinfreiheit, Sitzkomfort, Entspannung und Vergnügen bei der Reise. Der Sitz im Flugzeug übernimmt wirklich die Funktion der Lounge, des Konferenzzimmers, des Büros, des Eßzimmers und des Schlafzimmers, und es ist ein und derselbe Sitz, ob wir nun dick, dünn, kurz, lang, europäisch, amerikanisch, asiatisch, männlich, weiblich, alt, jung, entspannt, fröhlich, traurig oder gestreßt sind. Auch hier haben wir wieder mit den gleichen Problemen zu tun. Es gibt weltweit insgesamt zwölf Sitzhersteller, die Sitze für Fluggesellschaften produzieren. Die Grundanforderungen beziehen sich auf das Gewicht, die Größe und die Sicherheitsbedingungen beim Fliegen. Ob das entspannte Verweilen während des Fluges damit gelöst ist, bleibt fraglich."[1]

Dieser Zwang des stundenlangen Sitzens führte schon früh, spätestens in der Nachkriegszeit dazu, diese Passagiersitze als bequeme Lounge-Sessel zu gestalten. Kopfstützen, ergonomische Polster für Beckenunterstützung, ausklappbare Armlehnen, je nach Positionierung im Flugzeug auch Fußstützen setzen sich früh durch. Im Dreieck von Hygiene, Bequemlichkeit und Funktionalität entstand so mit den Passagiersitzen eine ganz eigene Produktgattung. Mit Klettbändern fixierte Hinterkopf-Servietten sind da ebenso obligatorisch wie kleine „Funktionsgeheimnisse" unter den Abdeckklappen der Armlehnen. Am Anfang bestanden diese Bedienfelder lediglich aus einem integrierten Aschenbecher und einem Knopf zum Rufen der Stewardess, dann wurden Steckerbuchsen für Kopfhörer zum individuellen Musikempfang und Videobuchsen integriert und seit 1998 gibt es in der First Class und der Business Class „Inseat-Power" für Laptops, seit 2004 sogar Internet-Zugang, genannt Fly-Net. Bei Interkontinentalflügen sind entweder aus der Armlehne herausklappbare TV-Flachbildschirme oder in die Rückseite des Vordersitzes integrierte Flachbildschirme für das individuelle Betrachten von Filmen an Bord inzwischen ebenso selbstverständlich wie die in den 1970er Jahren übliche gemeinsame Filmbetrachtung über eine aufgespannte Leinwand.

„Für die elektronische Unterhaltung an Bord ist jedem Passagier das ‚Personal Video System' zugänglich. Die Passagiere der Lufthansa können, jedenfalls in der ersten Klasse, ihr eigenes Unterhaltungsprogramm aus einer Auswahl von fünfzehn Filmtiteln, die in fünfzehn Sprachversionen zur Verfügung stehen, zusammenstellen, ergänzt durch ein umfangreiches Audioprogramm."[2]

Auch die entweder aus den Lehnen herausschwenkbaren oder von der Rückseite des Vordersitzes herunterklappbaren Tablare zur Aufnahme der Speisen und Getränke tragen zum Eindruck eines durchdachten Multifunktionsmöbels bei. Zudem muss ein solcher Passagiersitz auch Turbulenzen, Druckabfällen und hygienischen Extremsituationen gewachsen sein: Die Tüte im Staufach des Vordersitzes mag da als Hinweis gelten auf mögliche Unpässlichkeiten.

In der Regel seriell angeordnet, je nach Maschinengröße in waagrecht zur Gangrichtung angeordneten Reihen montiert, sollen und müssen diese Sitze den Spagat zwischen Sicherheit, Ergonomie und Ökonomie meistern. Die kleineren Maschinen haben zwei 3er-Reihen, also nur einen Mittelgang. Ihre Fachbezeichnung lautet deshalb „Narrow-Body Aircraft". Die größeren „Wide-Body Aircraft" mit zwei Gängen haben in der Economy Class unterschiedliche Sitzkonfigurationen: die Boeing 747 3-4-3, die DC10 2-5-2 (von 1973 bis Anfang 1994), die Airbusse A300/310/340 jeweils 2-4-2. Die Sitze sind in aller Regel schwebend auf Tragrohren montiert und bieten so zusätzlich unter der

Christian Steguweit, until 1994 director of brand management and corporate identity at Lufthansa, emphasized:

"Travel, particularly on long-haul flights, is reduced to one very essential function – seating comfort. In their advertising almost all airlines promise legroom, comfortable seating, relaxation and diversion during the trip. The airplane seat actually assumes the function of the lounge, conference room, office, dining room and bedroom and is always exactly the same seat, no matter whether we are fat or slender, short or tall, European, American, Asian, male, female, old, young, relaxed, happy, sad or stressed out. Here, too, we are confronted with the same problems. There is a total of twelve companies world-wide which manufacture aircraft seats. The basic requirements are referenced to weight, size and safety demands during flight. It remains questionable whether the design is sufficient to turn the flight into a relaxed journey."[1]

The requirement for passengers to remain seated for hours on end led quite early – during the post-war period at the latest – to the need to make these passenger seats as comfortable as lounge-type seating. Headrests, upholstery ergonomically designed for pelvic support, foldaway armrests and – depending on the seat's location within the aircraft – footrests quickly became commonplace. Within the design triangle defined by hygiene, comfort and function, the passenger seat emerged as a quite distinctive type of product. Headrest antimacassars secured with hook-and-loop tape are just as obligatory as those little "functional secrets" tucked away below the armrest covers. Initially, these miniature control panels included only an integrated ashtray and a stewardess call button. Headphone jacks for individual audio programming and video jacks were added later and, since 1998, First and Business Class travelers have had the convenience of "inseat power" for laptops. Internet access using the so-called Fly-Net has been available since 2004. On intercontinental flights individual, flat television screens – either integrated into the back of the seat in front or lifted out of the armrest – have in the meantime become as commonplace as watching movies together on a pull-down screen had been in the 1970s.

"For inflight entertainment, each passenger is offered a … Personal Video … As in the Lufthansa First Class cabins, passengers can create their own entertainment programs, drawing from 15 movie titles, in up to 15 different languages, plus a range of music tapes."[2]

The tray tables for meals and beverages – either swung up and out of the armrest or down from the back of the seat in front – contribute to the impression that airline seating is indeed well thought out, multifunctional furniture. Moreover, a passenger seat such as this has to be able to withstand turbulence, pressure loss and extreme hygienic situations. The paper bag tucked discretely into the seat pocket in front of the passenger suggests the types of mishaps that have to be dealt with.

Usually arranged in a serial configuration and – depending upon the carrier and the interior layout – in rows which are at right angles to the aisles, these seats have to give equal consideration to safety, ergonomics and economy. The smaller airframes have three seats on either side of a single center aisle. These are referred to as "narrow-body aircraft." The larger "wide-body aircraft" with two aisles in Economy Class exhibit a variety of seating configurations: 3-4-3 in the Boeing 747, 2-5-2 in the DC10 (from 1973 to the beginning of 1994) and 2-4-2 in the Airbus 300/310/340 models. As a rule these seats are designed to "float" on elevated carrier rails and thus not only offer personal storage space below the seat in front but also optimize cleaning. Distances between rows vary today according to the passenger classes: spaced for thrift in Economy Class, more generously dimensioned in Business Class with two meters of

Senator-Dienst, Erste Klasse, 1958
Senator service, First Class, 1958

Sitzfläche nicht nur individuellen Stauraum, sondern auch optimierte Reinigungsmöglichkeiten. Ihre Dimensionen variieren heute entsprechend den Flugklassen: ökonomisch aufgeteilt in der Economy Class, großzügiger dimensioniert für die Business Class mit zwei Meter Liegefläche auf 1,50 m Sitzabstand und in luxuriösen Sesselmaßen für die First Class. Diese hieß bei der Lufthansa jahrzehntelang „Senator-Klasse" und hielt als Kennzeichen von Anfang an für jeden Erste-Klasse-Passagier eine Rose bereit. Entsprechende Werbung für „luxurierende Beinfreiheit" zwischen den einzelnen Sitzreihen ist ein ambitioniertes Versprechen. Denn die Evidenz des ökonomischen Argumentes, in einer Maschine etwas enger positioniert 300 Sitzplätze oder eben etwas großzügiger lediglich 230 unterzubringen, ist mit Komfortüberlegungen nur schwer zu konterkarieren. Natürlich sind auch bei den Passagiersitzen Farb- und Formgestaltung wichtig und wahrnehmungsbestimmend. Aber eben auch die Funktionalität etwa der Sauerstoffmasken, die im Falle eines gravierenden Druckabfalles aus den über den Sitzen befindlichen Deckenequipments automatisch herunterfallen. Sie sind mit den Sitzplätzen ebenso in einem einheitlichen Entwurfs- und Gestaltungskonzept verbunden wie die Ruftasten oder die individuell regelbaren Sitzplatzbeleuchtungen über Kopf.

All dies sind seit Jahrzehnten gewachsene Standardniveaus im Passagiersitzangebot. Umso interessanter ist es, die Sitze der Lufthansa unter diesem engen Bedingungsgerüst daraufhin zu betrachten, ob es überhaupt noch gestalterische Freiheiten, Varianten oder gar Überraschungen gibt oder geben kann.

1957 wurde in der First- und der De-Luxe-Klasse der Lockheed Super Constellation der „Comforette"-Sessel eingeführt. Er hatte die Anmutung und Bequemlichkeit des Lounge Chairs von Charles Eames, der ein Jahr vorher entstanden war. Mit neigbarem Lehnenteil, volumigem Kopfkissen und ausklappbarer Fußstütze wurde der Sessel knapp über dem Boden montiert, so dass er keinerlei Stauraum unter dem Sitz bieten konnte. Von der gesamten Anmutung her war er noch mehr ein Wohnraum- als ein Flugzeugmöbel, noch unbeeinflusst von den ergonomischen und funktionalen Sachzwängen der letzten drei Jahrzehnte, die etwa eine Schockresorptionsfähigkeit der Materialien, aus denen die Sitzrahmen gefertigt werden, durch Tests vorschreiben. Diese sind heute unabdingbar bei der Entwicklung neuer Sitze und Voraussetzung für ihre Zulassung. Der Name „Comforette" ist ebenso den 1950er Jahren verpflichtet wie etwa jener des Staubsaugers „Vampyrette" aus der gleichen Zeit. Die Beinauflage sah etwas stakelig aus, gefertigt aus dünn dimensionierten Metallrohren. Zeitgenössische Bilder zeigen schlafende Passagiere dicht an dicht in einer Super Constellation, ohne Sicht- und Schallschutz untereinander: Das hat den etwas schlanken Charme einer Jugendherberge. Die mütterlich besorgte Stewardess wirkt denn auch eher wie eine Krankenschwester. Aber schon der „Comforette"-Sessel der First Class hatte in den Armlehnen integrierte Rufsysteme für die Flugbegleiter. Im Rückblick betont auch der Kulturwissenschaftler Jochen Eisenbrand:

„Betrachtet man Flugzeugsitze der Economy Class der 1960er und 1970er Jahre, so unterscheiden sie sich äußerlich kaum von heutigen Modellen. Wesentliche Neuerungen betrafen seitdem denn auch weniger die Sitzkonstruktion an sich als die Erweiterung und Verbesserung des Inflight-Entertainment-Systems. Dessen Bedienelemente wie Lautsprecherregler und Senderwahl, der Kopfhöreranschluss und die nötige Elektronik wurden in die Armlehnen integriert. Erst vor einigen Jahren sind auch LCD-Bildschirme in den Rückenlehnen üblich geworden, um von den Unannehmlichkeiten des Sitzens auf langen Flügen abzulenken."[3]

Flugzeugsitz „Comforette", ab 1957, First Class der Lockheed Super Constellation
"Comforette" aircraft seat, as of 1957, First Class section in the Lockheed Super Constellation

Bis weit in die 1970er Jahre hinein wurden die verschiedenen Sitzformen für die Lufthansa-Maschinen von den verschiedenen Herstellern übernommen. Lediglich die

Flugzeugsitze "Comforette" der Senator Class, Schlafposition, Lockheed Super Constellation, ab 1957
"Comforette" aircraft seat, Senator Class section, sleeping position, in the Lockheed Super Constellation, as of 1957

reclining length at seat pitch of 1.50 meters and luxurious armchair size for First Class. At Lufthansa this was known for decades as the Senator Class and from the very beginning provided as a distinguishing feature a rose for each First Class passenger. Advertising proclaiming "luxurious legroom" between the individual rows of seats is, when examined more closely, often an overly ambitious promise. It is difficult for comfort considerations to prevail over the persuasiveness of a simple economic fact: 300 seats can be installed when they are spaced more closely but only 230 with more generous legroom. Naturally, color and form are important in passenger seats and affect passenger perception. Certain functional aspects play a part here, as well. The overhead service panels reveal one aspect. They house the oxygen masks which, in case of a sudden pressure loss, are automatically released from their compartments above the seats. The individual reading lamps and, in some cases, the flight attendant call buttons are also mounted in these service units. It goes without saying that these service panels will have to be aligned exactly with the rows of seats.

All these are part of the standards for passenger seating which have developed over the decades. That makes it all the more interesting to consider Lufthansa's seating in light of this narrow corset of requirements to determine whether there is any design freedom at all and whether variants or even surprises have been or could be devised.

In 1957 the "Comforette" seat was introduced in the First and De-Luxe Classes in the Lockheed Super Constellation. It had the grace and comfort of the Charles Eames lounge chair, which had been created one year before. With its adjustable backrest, plump head cushion and foldaway footrests, the seat was mounted just a few inches off the floor, eliminating any stowage beneath the seat. The overall impression was more of living room furnishings than aircraft seating and was as yet not influenced by the ergonomic and functional requirements of the past three decades. One example here is mandatory testing to verify the shock absorbing capability of the materials from which

Kabine, First Class, Boeing 707, 1965
First Class cabin, Boeing 707, 1965

Flugzeugsitze, First Class, Boeing 747-100, um 1975
Aircraft seats, First Class, Boeing 747-100, about 1975

Stoffdessins der Sitze und weitere textile Ausstattungen wie etwa die Teppichböden wurden von der hauseigenen Designabteilung beigesteuert. 1972 waren die Sitze in der Tourist Class der Boeing 747 mit türkisem bzw. grünlichem Stoff bezogen, nun schon deutlich schwebender und nur noch mit einer „Kopfserviette“ bestückt. Die kunstlederummantelten Armlehnen waren aufklappbar und gaben so die Ruftasten frei. In der First Class steckte die berühmte Rose in jeder Lehne der orange bezogenen Sitze, in der Touristenklasse der Boeing 707 gab es blau bzw. rot bezogene Stoffsitze in tiefgezogenen Kunststoffschalen. Einer LH-Werbeanzeige von 1974 ist zu entnehmen, dass diese Sitze von einem italienischen Professor Lenci entworfen wurden. Als Besonderheit fällt auf, dass ein Ablagetablar zwischen die Armlehnen der Sitze fixiert werden konnte. Die um 1975 fotografierte Lounge im Oberdeck einer Boeing 747-130 der LH zeigt erneut die zeittypische Dominanz von Orange- und Brauntönen. Die Formen sind abgerundet, „gesoftet“. Erdnüsse, Chips, Olivensticker und orange Aschenbecher auf den weißen, quadratischen Couchtischen: Bar-Atmosphäre! Die Sitzbezüge bis 1982 kennzeichnet ein Farbkonzept grau-beiger, gelb-gelber und gelb-oranger Fischgrätstreifen. Die Filmvorführungen erfolgen über die damals üblichen großdimensionierten Decken-Projektoren. In den späten 1970er/frühen 1980er Jahren wurden die Sitze in der ersten Klasse muskulär gepolstert; sie wirkten beinahe wie englische Clubsessel. Die Rückenlehne des eierschalenfarben bezogenen Sitzes ist in drei waagrechte Wülste separiert, die volumigen klappbaren Armlehnen wirken wie hartkantige Primärformen und die fest miteinander verbundenen Doppelsitze sind auf grazilen Rundstäben ohne Fußstützen aufgeständert, was die Reinigung der Böden wesentlich erleichtert. Entworfen hatte diese bequemen, frühen „High-Touch“-Sitze der amerikanische Industriedesigner Walter Dorwin Teague gemeinsam mit den Industriedesignern der LH. Die Entwürfe ebenso wie die Materialmusterkarten zeigen Weiß, Orange, Braun, aber auch Hellgrau als Farbklangkonzept. Die Alu-Außenarmlehnen sind wie ein offener „C“-Buchstabe mon-

Flugzeugsitze, Economy Class, Boeing 747, 1972
Aircraft seats, Economy Class, Boeing 747, 1972

the frame is manufactured. Today such testing is an absolute necessity in the development of new seats and a prerequisite for their approval. The name "Comforette" is just as much as child of the 1950s as, for example, that of the "Vampyrette" vacuum cleaner of the same era. The leg rest looked a bit spindly, being made of thin metal tubing. Contemporary photographs show sleeping passengers cheek by jowl in a Super Constellation, without the luxury of any visual or sound shielding. The entire situation radiated the Spartan charm of a youth hostel. And the mothering stewardess seemed more like a hospital nurse. But the "Comforette" seat in First Class already had a flight attendant call system integrated into the armrest. In a retrospective view, cultural analyst Jochen Eisenbrand also emphasizes:

"If we examine Economy Class aircraft seats in the 1960s and 1970s we find that externally they differ hardly at all from today's models. Important innovations since that time have affected less the engineering of the seats and more the expansion and enhancement of the inflight entertainment system. The control elements, including volume control and channel selection, headphone jack and the associated electronics, were integrated into the armrests. It was not until just a few years ago that LCD screens in the backrests became common, helping to divert attention from the discomfort of sitting on long flights."[3]

It was until well into the 1970s that the various seating manufacturers determined the shape of the seats for Lufthansa's jetliners. The company's own design department merely provided the designs for the upholstery materials and other textile appointments such as carpets. In 1972 the Tourist Class seats in the Boeing 747 were upholstered in turquoise or a greenish material, looked more as though they were "suspended in mid-air" and were fitted only with a kind of antimacassar. The tops of the armrests, covered with artificial leather, were hinged and could be raised to give access to the call buttons. In First Class the famous rose was placed in one armrest at each orange-upholstered seat, while in Tourist Class in the Boeing 707 there were blue and red upholstered seats in deep-drawn plastic shells. A Lufthansa advertisement in 1974 proclaimed that this seat had been designed by the Italian Professor Lenci. One special feature was a tray table which could be mounted between the armrests. The lounge on the upper deck of a Boeing 747-130 operated by Lufthansa, photographed around 1975, shows once again the dominance of the orange and brown shades so typical of that period. The shapes are rounded and the edges softened. Peanuts, potato chips, olive picks and orange ashtrays on the low, square, white coffee tables impart a pub-like atmosphere. The upholstery for the seats through 1982 reflected a color scheme including gray and beige, yellow on yellow, and yellow and orange herringbone stripes. Movies were shown by way of the bulky ceiling projectors common at that time. In the late 1970s and early 1980s the seats in First Class could clearly be characterized as "overstuffed" and reminiscent of English club chairs. The backrest, upholstered in an eggshell colored fabric, is divided horizontally into three sections. The bulky, hinged armrests remind the viewer of hard-edged, basic geometric shapes and the pairs of seats, joined permanently one with the other, are mounted on rather delicate round profiles without footrests, which significantly simplifies floor cleaning. This comfortable seat, an early high-touch product, was crafted by American industrial designer Walter Dorwin Teague in collaboration with the industrial designers at Lufthansa. The sketches and the swatch books for the fabrics show white, orange, brown – but also light gray – in a harmonious color scheme. The aluminium outboard armrests are mounted in such a way as to suggest an open "C." The rigid shell and its integrated upholstery give an impression of both technical precision and comfort. But with the dark floor covering and the dark brown finish on the

Flugzeugsitze, Tourist Class, Boeing 707, um 1975
Aircraft seats, Tourist Class, Boeing 707, about 1975

Werbeanzeige mit Flugzeugsitzen, 1974
Advertisement featuring aircraft seats, 1974

Schnittisometrie einer Boeing 747-100, 1970er Jahre
Isometric section of a Boeing 747-100, 1970s

Lounge im Oberdeck, Boeing 747-130, um 1975
Upper deck lounge, Boeing 747-130, about 1975

partitions, the interior of this cabin shows something which is typically American: solid and distinguished, rather like an executive suite furnished by Florence Knoll. These seats were introduced in Business Class at the same time but the backrests, in dark brown here, are far more elegant as they form a smooth surface following body contours. This is just as true for the Tourist Class seats of the same period, showing red and gray or brown and beige upholstery. These three types of aircraft seat were used until 1987. Beginning in May 1980 the Senator Sleeper developed together with Weber Air-craft were used on all Boeing 747 flights. Here the backrests could be tilted by as much as 60 degrees so that passenger could, as Lufthansa advertising put it: "Reach your destination reclining in comfort."

As of 1988 the Zurich-based Zintzmeyer & Lux agency, which had been commissioned with a relaunch of the entire Lufthansa corporate identity, harmonized the appearance for the seats in all the three classes: F (First Class), C (Business Class) and M (Economy Class). For the first time there was a uniform upholstery concept for all three classes: silver against gray for the F class, blue against gray for the C class and yellow against gray for the M class. The tray tables in the F and C classes were lifted out of the armrests while in the M class they were fastened to the back of the seat in front. In the C class there was now in-seat video with the following instructions printed on the frame: "Please stow for take-off and landing." In case a take-off should be aborted, passengers were to be able to reach the exits quickly without becoming entangled in the monitors. The continuity of the aircraft seating design was further emphasized with special components such as escutcheons – the front caps on the armrests – made of stainless steel in every class. These seats were flown through 1996 even though the design activities at Zintzmeyer & Lux had drawn to a close in 1990. In addition, the seats had quite obviously become "sitting machines" with fold-away footrests and even more inflight electronics integrated into the armrests. The patterns for the textiles, with reserved, textured, diagonal stripes in the Lufthansa colors, made a clear reference to the airline's corporate design. The First Class seat in this series was the first electronic seat. Even a "swivel" seat, which could be rotated to form groups of 4 or 8, was built for the Boeing 747-400. Although publicized, it was never put into service.

Thus in 1992/93 Lufthansa's in-house designers worked with the upholstery specialists at Keiper-Recaro to develop variable-width seats for the newly introduced Lufthansa Express Business Class service inside Germany. An enthusiastic Tilman Jens also wrote in 1994:

"A few simple actions suffice and the six-abreast configuration dreaded by frequent fliers becomes an array for five. This puts an end to the days in which simply reading a newspaper in a jetliner was a serious invasion of the neighbor's personal space. The three seats on the left-hand side – as I read in a brochure in the waiting area – can be converted to a pair by shifting the armrest. There are still three seats on the right-hand side. They can, however, be pulled out by about 3 cm, rather like an accordion, whenever that row is to lay claim to the 'business' title ... Anyone who has a chance to examine Lufthansa's new German domestic Business Class invariably sings its praise. Lufthansa's strategists certainly landed a trail-blazing coup."[4]

These seats, dubbed "CVS" (convertible seat) by Lufthansa, made it possible, in rows 16 to 29, depending on the airframe, to convert five Business Class seats into six Economy seats. Consequently capacity could be varied depending on demand – either more or fewer but wider seats. That is why the article cited here also posed the question: "A classless society above the clouds?" After initial difficulties in the years 1992/93, Lufthansa's entire narrow-body fleet and a part of the A300 fleet was, as of 1995, outfitted in

Stoffmuster für die First Class und die Economy Class, Mitte 1970er Jahre, Hersteller: Lantal (ehem. Langenthal)
Fabric samples for First Class and Economy Class, mid-1970s, manufacturer: Lantal (formerly Langenthal)

Werbeanzeige mit dem Senator Sleeper, Boeing 747, 1980
Advertisement featuring the Senator Sleeper, Boeing 747, 1980

Flugzeugsitze, Economy Class, Airbus A300-B2, Hersteller: Recaro Aircraft Seating, Ende der 1970er Jahre
Aircraft seats, Economy Class, Airbus A300-B2, manufacturer: Recaro Aircraft Seating, end of 1970s

Zintzmeyer & Lux: Flugzeugsitze, First Class, 1990–1996
Zintzmeyer & Lux: Aircraft seats, First Class, 1990–1996

Zintzmeyer & Lux: Flugzeugsitze, First Class, 1990–1996
Zintzmeyer & Lux: Aircraft seats, First Class, 1990–1996

tiert. Gemeinsam mit der Hartschale, in der die Polsterung integriert ist, entsteht neben dem Eindruck technischer Präzision ebenso der Eindruck von Komfort. Aber durch den dunklen Boden und die dunkelbraune Trennwandverkleidung hat dieses Kabineninterieur auch etwas typisch Amerikanisches: schwer, gediegen, wie eine von Florence Knoll eingerichtete Vorstandsetage. Bei den gleichzeitig in der Business Class eingesetzten Sitzen sind die nun dunkelbraunen Rückenlehnen wesentlich eleganter, weil sie eine glatte Körperschale bilden. Dies gilt ebenso für die zeitgleichen Passagiersitze der Tourist Class, die rot-graue bzw. braun-beige Polsterungen aufwiesen. Diese drei Flugsitztypen waren bis 1987 im Einsatz. Ab Mai 1980 kamen auf allen Boeing-747-Flügen die gemeinsam mit Weber Aircraft entwickelten Senator Sleeper zum Einsatz, deren Rückenlehnen sich bis zu 60° verstellen ließen. Wie eine LH-Werbung formulierte: „Bequem liegend ans Ziel kommen".

Ab 1988 harmonisierte die Zürcher Agentur Zintzmeyer & Lux, die den Auftrag für ein Relaunch des gesamten Lufthansa-Erscheinungsbildes erhalten hatte, das Erscheinungsbild der Sitze für die drei Klassen F (First Class), C (Business Class) und M (Economy Class). Zum ersten Mal gibt es ein durchgängiges Sitzbezugdekor für alle Klassen: silber auf grau für die First Class, blau auf grau für die C-Klasse und gelb auf grau für die M-Klasse. Die Tablare zum Speisen waren bei der F- und C-Klasse aus den Armlehnen herauszuklappen, bei der M-Klasse an der Rücklehne des Vordersitzes befestigt. In der C-Klasse gibt es jetzt „Inseat-Video", mit dem auf der Vorderseite des Bildschirmrahmens aufgedruckten Hinweis: „Bitte für Start und Landung verstauen", denn im Falle eines Startabbruchs sollen die Passagiere schnell zu den Ausgängen kommen, ohne an den Monitoren „hängenzubleiben". Die „Durchgängigkeit" des Flugsitzdesigns wurde durch spezielle Bauteile wie Escutcheons aus V2A-Stahl, das sind die Frontkappen der Armlehnen, in allen Klassen verdeutlicht. Diese Sitze wurden zwar bis 1996 geflogen: die Entwurfstätigkeit von Zintzmeyer & Lux war allerdings bereits 1990 beendet. Zudem wurden die Sitze nun deutlich zu „Sitzmaschinen" mit ausklappbaren Fußstützen, nochmals vermehrter, in den Armlehnen integrierter Inflight-Elektronik und Stoffdessins, die mit diagonal strukturierten Streifen in den Lufthansa-Farben deutlich den Bezug zum „Corporate Design" der Fluggesellschaft herstellen. Der First-Class-Sitz dieser Baureihe war der erste elektronische Sitz. Sogar ein sogenannter „Swivel"-Sitz, der sich zu einer 4-er und 8-er Konstellation drehen ließ, wurde für die Boeing B747-400 gebaut und auch beworben. Zum Einsatz kam er allerdings nicht.

Und so entwickelten 1992/1993 für die damals neue innerdeutsche Business Class der Lufthansa, die LH-Express, hauseigene Entwerfer zusammen mit den schwäbischen Polsterspezialisten der Firma Keiper-Recaro Flugsessel, die in der Breite variabel sind.

„Wenige Handgriffe genügen", wie Tilman Jens 1994 enthusiasmiert schrieb, „und aus der von Vielfliegern gefürchteten Sechserreihe wird eine Phalanx für Fünfe. Vorbei sollen sie sein, die Tage, in denen schon die Zeitungslektüre im Jet einer Freiheitsberaubung des Nebensitzers gleichkam. Die drei Sitze auf der linken Seite, so durfte ich schon der im Warteraum ausliegenden Broschüre entnehmen, lassen sich durch Verschieben der Armlehne zu einer Zweierreihe umfunktionieren. Rechts bleibt es bei drei Plätzen, die man freilich um jeweils rund drei cm wie ein Akkordeon ausziehen kann, wenn sie das Prädikat ‚Business' in Anspruch nehmen sollen" und weiter: „Wer die neue innerdeutsche Business-Klasse der Lufthansa begutachten darf, gerät unweigerlich ins Schwärmen. Da ist den Strategen des Luftfahrt-Unternehmens wahrlich ein zukunftsweisender Coup gelungen."[4]

Diese von der Lufthansa „CVS" (Convertible Seat) genannten Sitze erlaubten, in den Reihen 16-29, je nach Flugzeug-Typ der Inlandsflieger fünf Business-Class-Sitze in

Zintzmeyer & Lux: Flugzeugsitze, Business-Class-Kabine, 1990–1996
Zintzmeyer & Lux: Aircraft seats, Business Class cabin, 1990–1996

the forward section (about 80%) with the convertible seats made by Recaro Aircraft Seating. Once again Lufthansa's own design department created the leather upholstery for the CVS seats. These are in the distinguished "business gray" which has been used by Lufthansa ever since the Zintzmeyer & Lux color scheme was adopted. This is accentuated with yellow piping.

It was this very same Tilman Jens, journalist and notorious frequent flier, who commented with a touch of irony on the new line of seats developed as of 1996 by Frog Design, Hartmut Esslinger, the chief and creative genius at the multinational design firm, had dubbed the look "Retro-Futurism" with great public appeal.

"Months of joyous anticipation of discovering – as promised in a lavishly designed brochure from the crane's home base – a new dimension in flying on intercontinental routes: 'Relaxation for body and spirit.' How often have I been tortured – not only on domestic hops, but also on intra-European flights – by Lufthansa seating mounted at absurdly close distances, in seats which promise to cramp the extremities of – and I say this with slight understatement – any aspiring basketball player. I mutter quiet resolutions to at last enroll in yoga lessons!"[5]

Taking part in a seating test at a presentation for travel agency owners in a hangar decked out in a circus-like atmosphere on the Lufthansa grounds, the editor initially thinks:

"I'm determined to find these new seats horrible but then lounge in the Business-Class seat on exhibit and soon abandon all my evil intentions. The seat – there is no way around it – is sensationally comfortable, the distance to the seat in front is a full 122 cm, 50% more than on my last trip to New York. The tilt angle has also been significantly increased. Storage space is in abundance and, to top it all off, there is a socket for the laptop computer. Then I take a look at First Class. On this particular evening it was set up directly opposite. From this moment on the passenger can lie absolutely flat. There is a full 2.34 meters to the seat in front."[6]

Inseat-Video, Business Class, 1990–2003
Inseat video, Business Class, 1990–2003

Convertible Seats (CVS), Business Class, Einsatz nur im Kontinentalverkehr, ab 1995
Convertible seats (CVS), Business Class, used only on Continental routes, as of 1995

sechs Economy-Sitze zu verwandeln. Je nach Buchungslage konnte also das Angebot variieren: entweder mehr oder weniger, dafür breitere Sitze. Deshalb fragt der zitierte Artikel auch: „Klassenlosigkeit über den Wolken?“ Nach anfänglichen Schwierigkeiten in den Jahren 1992/93 wurde ab 1995 die gesamte „Narrow-Body-Flotte“ der Lufthansa und ein Teil der A300-Flotte im vorderen Bereich (ca. 80%) mit den „Convertible Seats“ von Recaro Aircraft Seating ausgestattet.

Erneut entwickelte die firmeneigene Designabteilung der Lufthansa für CVS-Sitze Lederbezüge. Diese sind im seit Zintzmeyer & Lux zur Farbpalette der Lufthansa gehörenden vornehmen „Geschäftsgrau“ gehalten: mit dezent sich absetzenden gelben Paspelkanten.

Der nämliche Tilman Jens, Journalist und notorischer Vielflieger, kommentierte auch leicht ironisch die ab 1996 entwickelte neue Sitzlinie von Frog Design, die Hartmut Esslinger, Chef und Kopf der multinationalen Designschmiede, publikumswirksam „Retro-Futurismus“ nannte:

„Vorfreude seit Monaten, gilt es doch, wie eine aufwendig gestaltete Broschüre aus der Heimstatt des Kranichs anzeigt, auf interkontinentalen Strecken eine neue Dimension des Fliegens zu entdecken: Entspannung für Körper und Geist. Wie oft habe ich mich auf innerdeutschen, aber auch europaverbindenden Flügen in enges, in verflucht dichtem Abstand montiertes LH-Gestühl gequält, in Sitze, die jedem, der, sagen wir es mit leisem Understatement: über preußisches Gardemaß verfügt, die Glieder zu stauchen versprechen. Stumme Appelle, doch endlich Yoga-Unterricht zu nehmen!“[5]

Beim Probesitzen bei einer Sitzpräsentation für Reisebüroinhaber in einem zirzensisch umgestalteten Hangar auf dem Lufthansa-Gelände denkt der Redakteur zunächst:

„Ich beschließe, die neuen Sessel schrecklich zu finden, lümmele mich in den ausgestellten Business-Class-Platz und gebe bald alle bösen Vorsätze auf. Der Sitz – es hilft nichts – ist sensationell bequem, der Abstand zum Vordermann beträgt ganze

„Ich beschließe, die neuen Sessel schrecklich zu finden, lümmele mich in den ausgestellten Business-Class-Platz und gebe bald alle bösen Vorsätze auf. Der Sitz – es hilft nichts – ist sensationell bequem.“

“I’m determined to find these new seats horrible but then lounge in the Business Class seat on exhibit and soon abandon all my evil intentions. The seat – there is no way around it – is sensationally comfortable.”

**Tilman Jens, Journalist, 1998**

Frog Design Flugzeugsitze, First Class, ab 1998; Hersteller: Weber Aircraft
Frog Design: Aircraft seats, First Class, as of 1998; manufacturer: Weber Aircraft

Frog Design: Flugzeugsitze, Business Class, ab 1998; Hersteller: Weber Aircraft
Frog Design: Aircraft seats, Business Class, as of 1998; manufacturer: Weber Aircraft

Frog Design: Flugzeugsitze, Economy Class, ab 1998; Hersteller: Recaro Aircraft Seating
Frog Design: Aircraft seats, Economy Class, as of 1998; manufacturer: Recaro Aircraft Seating

122 cm, 50 % mehr als auf meinem letzten Trip nach New York. Auch der Neigungswinkel ist spürbar erhöht. Stauflächen in Hülle und Fülle, überdies eine Adapterdose für den Laptop. Dann nehme ich mir die erste Klasse vor. Sie ist an diesem Abend gleich gegenüber aufgebaut. Hier kann man fortan flach liegen. Bis zum Vordersitz sind es 2,34 m!“[6]

Esslingers Sitze huldigen einer Wellblech-Renaissance der legendären Ju 52, weisen aber auch jene additive Formenstapelung auf, die seit der Postmoderne und der Memphis-Bewegung in den 1980er Jahren als „Mikroarchitektur“ den prinzipiellen Entwurfskanon erweitert hat. Und so haben die Frog-Designer – an dem LH-Projekt waren die Frog-Teams von Peter Weber und Laurenz Schaffer, Altensteig, sowie Sonja Schiefer, Sunnyvale, beteiligt – gerade die soften Volumen der Sitzflächen und Kopfstützen demonstrativ aufgespreizt in blütenartige Szenarien. Bei der ersten Klasse thront über einem preußischblauen Sitz – eine Farbe, die Hartmut Esslinger werbewirksam zum Symbol der Arbeitsethik verklärt – ein Triumvirat von drei silbergrauen Kopfpolstern, die das Schlummern in verschiedenen Körperhaltungen optimieren sollen. Die Lehnen dieser Erste-Klasse-Sitze im gewellten Alu-Touch mit obligater Rose im Väschen, die weiland schon die Senator-Klasse begleitete, sind muskulös wie eine Werkbank, aber mit allen bordelektronischen Raffinessen ausgestattet:

„Wer selbst auf den Thron steigen möchte, muß – etwa für das Rückflugticket Frankfurt am Main–San Francisco – gut 13.000 DM investieren. Dafür ist in die Armlehne eine Vase für die rote Senator-Rose eingelassen, an der Seite eine Schuhbox, und neben dem Kopfkissen suggeriert ein herausklappbares Trennblech die Illusion von Privatheit am Rand der Stratosphäre ... Die Luxus-Menüs werden auf Abruf serviert, Zeitpunkt und Abfolge der Gänge bestimmt allein der Passagier, genauso wie das Bordunterhaltungsprogramm, das in der First Class nun Personal Video System heißt. In jedem Sitz steckt ein Recorder ... auch so läßt sich Freiheit begreifen. Als schiere Sitzgelegenheit sind die Flug-Sessel vortrefflich.“[7]

Ursprüngliche Vorstudien für die First-Class- und Business-Class-Sitze sahen ausziehbare Hauben zum Licht- und Schallschutz vor. Die Lufthansa beschloss aber nach durchaus anfänglicher Begeisterung für dieses Konzept aus Gründen der Sicherheit, diese halbrunden Hauben mit der Anmutung einer Kinderwagenabdeckung nicht zu realisieren. Auch das Schlafen bzw. Relaxen auf Langstreckenflügen ist nun wesentlich entspannter und bequemer als – sagen wir – noch zu Zeiten der Super Constellation. In der Business-Klasse sind nun die drei Kopfpolster in Lufthansa-Blau gehalten, während der Sitz im vornehmen Grau sich hauchdünne weisse Striche leistet. Auch hier wieder fallen die gewellten Alublech-Lehnen auf, die in ihren Nierentischschwüngen eine deutliche 1950er-Jahre-, aber eben auch eine „Tante Ju“-Nostalgie vermitteln. Die Economy-Klasse schließlich weist melonengelbe Kopfstützen, ebenso melonengelbe Kissen und Streifen im grauen Sitzbezug auf: nirgends auch nur ein Anflug des Lufthansa-Blaus. An der Trennwand des Fliegers zur nächsten Klasse das obligate Wellblech und der Kranich im Kreis: nun plastisch und in gelb. Die Wellen und ineinander geschobenen Radien der Sitzpolster und -lehnen vermögen Esslingers Entwurfscredo „forms follows emotion“, das er von Anfang an dem funktionalistischen Motto „form follows function“ gegenüberstellte, ideal zu vermitteln. Bis heute werden diese Sitze in der Boeing 747 und im Airbus A340-300 verwendet. Mit einer identischen Strategie hatte Esslinger mit seinen Frog-Teams bereits kurz vorher am Boden die Counter und Lounges der Lufthansa verändert.

Ab 2004 kommen die neuen Sitze des Londoner Designbüros Priestman Goode in der Business Class auf Langstreckenflügen zum Einsatz, die in kühler silber-blauer

Esslinger's seats pay reverence to the renaissance of the legendary JU 52 in corrugated sheet metal. They also show the additive use of shapes sometimes – as of the post-modern and the Memphis movement in the 1980s – referred to as "micro-architecture." This technique expanded the basic range of design instruments. And thus the Frog designers had demonstratively spread out the soft volumes of the seat surfaces and headrests to create flower-like scenarios.

In First Class, atop a Prussian blue seat – a color which Hartmut Esslinger has transfigured with good advertising effect to a symbol of the work ethic – is found a triumvirate of three silver-gray head cushions which are intended to optimize napping in various postures. The armrests in the First Class seats with their corrugated aluminum touch and the obligatory rose in the vase, which in days past marked the Senator Class, are as brawny as a workbench but are equipped with every electronic inflight refinement:

"Anyone who wishes to mount the throne will have to invest a good DM 13,000 – the price of a round-trip ticket between Frankfurt and San Francisco, for instance. But at that price there is a vase recessed into the armrest for the red Senator rose, there is a shoe compartment at the side while a folding panel next to the head cushion suggests an illusion of privacy at the edge of the stratosphere … The luxury menus are served on request – the passenger alone determines the time and sequence of the courses, just as with the inflight entertainment program which in First Class is now called the Personal Video System. There is a video player tucked away in every seat … and that is one way in which the essence of freedom can be grasped. As seating, pure and simple, these airline seats are superb."[7]

The original preliminary studies for the First- and Business Class seats provided for telescoping bonnets to shield against light and noise. After initial enthusiasm for this concept Lufthansa ultimately decided, in the interest of safety, not to adopt these semi-cylindrical bonnets vaguely reminiscent of a baby carriage hood. Sleeping and relaxing on long-haul flights is nonetheless far more stress-free and comfortable than – let us say – in the era of the Super Constellation. In Business Class the three head cushions are now in Lufthansa blue, while the seat, in elegant gray, is graced by white pinstripes. Here again the corrugated aluminum armrests catch the eye and, with their lines reminding us of the famous kidney-shaped table, they impart a clear sense of nostalgia associated with the 1950s and even with the "Auntie Ju." The Economy Class has melon yellow headrests and the same melon yellow for the supplementary cushions and the stripes in the gray upholstery. There's not a hint of Lufthansa blue to be seen. On the partitions separating the classes there is the obligatory corrugated sheet metal and the crane within a circle, now in relief and in yellow. The corrugations and the intersecting radii of the upholstered seats and backrests project ideally Esslinger's design credo – "form follows emotion" – which he from the very beginning contrasted with the function-alist "form follows function" motto. These seats have to date been installed in the Boeing 747 and the Airbus A340-300 aircraft. Following an identical strategy, Esslinger and his Frog teams had already modified the Lufthansa counters and lounges on the ground shortly prior to this.

As of 2004 the new seats developed by the Priestman Goode design office in London were put into service in Business Class on long-haul flights. With their cool silver and gray color scheme and a daring parabolic curve for the shell of the seat, reaching all the way down to the floor, they now embody absolutely the high-tech esthetics of an aerospace interior. These seats unfold a luxurious two meters of reclining length as soon as they are moved into the sleeping position. The structure of the seat's shell and

Frog Design: Inseat-Video, Business Class, ab 1998
Frog Design: Inseat video, Business Class, as of 1998

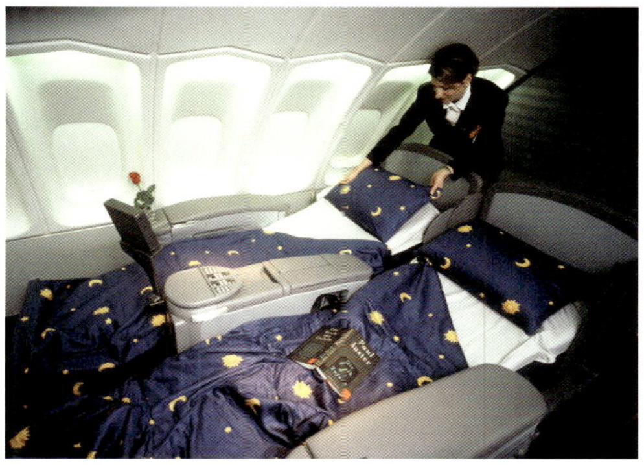

Frog Design: Flugzeugsitze, First Class, ab 1998
Frog Design: Aircraft seats, First Class, as of 1998

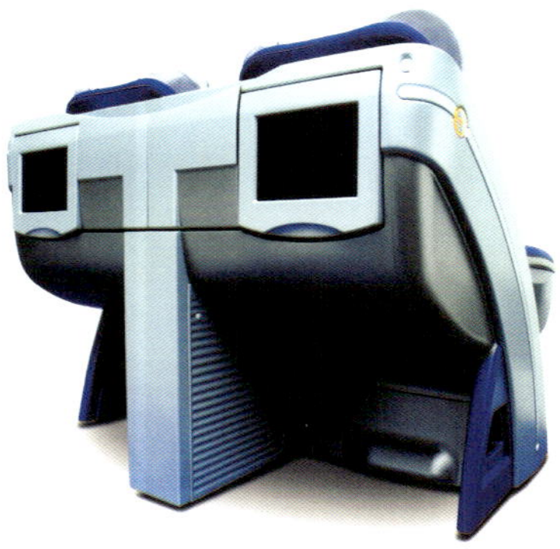

Priestman Goode: Flugzeugsitze, Business Class, ab 2004
Priestman Goode: Aircraft seats, Business Class, as of 2004

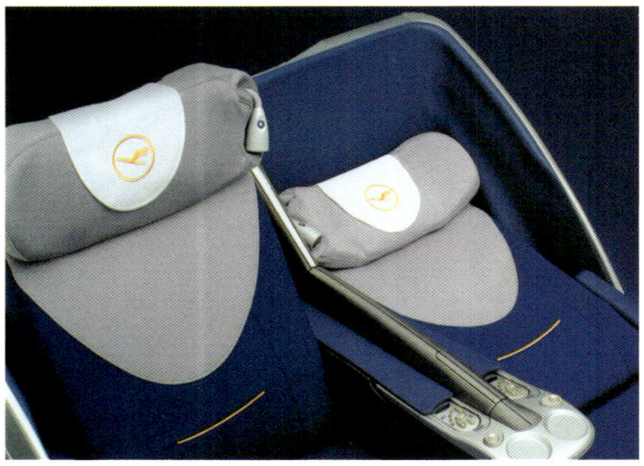

Priestman Goode: Flugzeugsitze, Business Class, Sitzlehnen, ab 2004
Priestman Goode: Aircraft seats, Business Class, seat backs, as of 2004

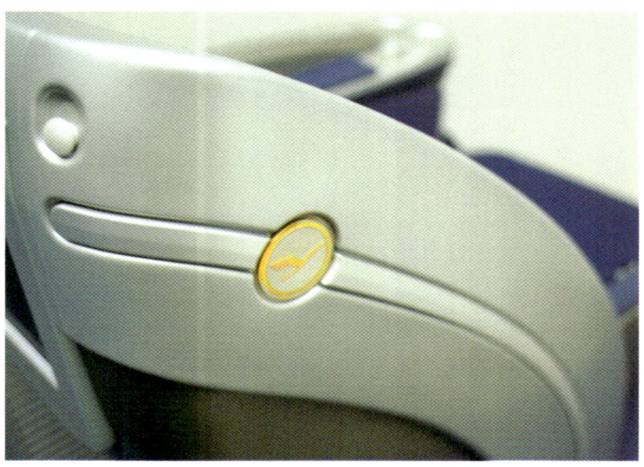

Priestman Goode: Flugzeugsitz, Seitenverkleidung, Kopfstütze und Sitzschale, ab 2004
Priestman Goode: Aircraft seat, side cladding, headrests, seat shell, as of 2004

Farbgebung und in einem kühnen Parabelschwung der Sitzschale, der bis zum Boden reicht, nun endgültig die High-Tech-Ästhetik eines Space-Interieurs vermitteln. Bei diesen Sitzen sind, sobald man sie zur Schlafposition ausfährt, luxuriöse 2 m Liegeflächen realisiert. Die leicht anmutende Konstruktion der Sitzschale und des Rahmens bildet einen auffälligen visuellen Fokus und lenkt die Aufmerksamkeit auf die vielen „Extras" des Sitzes. In die Sitzschale und den Bezug ist die Bildmarke des Kranichs im Kreis als deutliches CI Bekenntnis markant integriert. Die fest verbundenen Doppelsitze haben eine avancierte Videotechnik im Rückenteil. Die Formensprache der Sitze ist in beiden Klassen, Business und Economy, die gleiche. Die Teppichböden sind nun blau gemustert, neue Decken und Kissen stammen ebenfalls von Priestman Goode. Bei den Rückenwänden wurde das Wellblech nun wesentlich zurückhaltender eingesetzt und ist nur noch als Zitat der Frog-Produktsprache vorhanden. Formal sind diese Sitze jener „Hard-Edge-Formgebung" verpflichtet, die auch die aktuellen Autoentwürfe – von Audi bis Renault – kennzeichnet. Das zum Flugsitz gehörende Unterhaltungs- und Informationsprogramm sieht neben einer Video-Fernbedienung mit über 20 Optionen auch ein sogenanntes „Gamepad" vor: Reflex auf den millionenfach verkauften „Gameboy" und seine Nachfolger von Nintendo.

Für den Airbus A340-600 entwickelte 2003 das Hamburger Büro Jochen Müller und Jens Romca Industrial Design neue Flugsitze, die das Einsteigen in die und das Aussteigen aus den Flugzeugen ebenso wie ihre Reinigung erleichtern. Es sind eher bescheidene Dimensionen, die sich da zeigen, aber sie sind wohl kaum dem Entwurfsbüro anzulasten. Obwohl die Ausführungsqualität der Sitze exzellent ist, ist auch hier die Beinfreiheit erneut ziemlich knapp ausgefallen. Bisher allerdings sind diese Sitze nicht bei der Lufthansa zum Einsatz gekommen.

Ebenso komplex wie die Passagiersitze sind innerhalb der Bordausstattungen der Flugzeuge die Nasszellen, also die Bordküchen, die sogenannten „Galleys", und die Toilettenräume. In den Maschinen für Inlands- und Europastrecken sind die Küchenblöcke meistens in der Querung von einer zur anderen Tür untergebracht. Auf kleinstem Stauraum müssen die Flugbegleiter auf minimierten Zubereitungsflächen die Getränke und Speisen vorbereiten, die dann, je nach Flugklasse, mit Servierwagen bzw. Trolleys zu den verschiedenen Sitzreihen gebracht werden. Fast immer sind diese Wände der Bordküchen aus Honey Comb und die Arbeitsflächen, Profile und Verschlussriegel aus ebenso leichtem wie widerstandsfähigem Aluminium gefertigt. Sie bestehen in der Regel aus übereinander gestapelten Containern, Schubladen und Ausziehfächern für die Getränke. Ihre nüchterne Funktionalität ist prinzipiell dekor- und ornamentlos. Ihre Formensprache erinnert an die kühle Professionalität von Großküchen, von denen offensichtlich ihre Ästhetik abgeleitet ist. Damit stehen die Bordküchen in scharfem Gegensatz zum emotional aufgeladenen Bordinterieur, mit dem die Fluggäste direkt in Kontakt kommen. Allenfalls die Servierwagen in der ersten Klasse strahlen eine gewisse formale Eleganz aus. Und dann natürlich die Bartresen in den Oberdecks der zweigeschossigen Interkontinentalflugzeuge wie etwa der Boeing 747, als eben noch nicht die größtmögliche Passagierzahl das Hauptkriterium für Effizienz war. Während etwa die Bordküche der Boeing 707 um 1959, in der bis zu 168 Mahlzeiten zubereitet werden konnten, noch die typischen Eisfarben des Jahrzehnts – Blassgelb, Himmelblau und Rosarot – aufweist, ist dann die Bordküche im Hauptdeck einer Boeing 747 um 1970 schon wesentlich größer und mit äußerster Rationalität konstruiert. Sie strahlt eine Anmutung aus, mit Temperaturmessgeräten zwischen den Containerfächern und eingelassenen Griffen, die deutlich an rationale Laborbedingungen denken lässt. Die Bordküchenausstattung einer Boeing 747 aus den frühen 1970er Jahren etwa neben einer

frame, with a feeling of lightness, forms a striking visual focus and draws attention to the many "extras" which the seat offers. The crane emblem in the circle is integrated prominently into the shell and the upholstery material as a clear commitment to corporate identity. The tandem seats integrate advanced video technology in the backrest. An identical design vernacular is used for the seats in both classes – business and economy. The floor coverings are now patterned in blue and new blankets and pillows are also a Priestman Goode product. The use of corrugated metal is far more restrained in the rear walls and is now present only as a citation of the Frog product language. At the formal level these seats are committed to the "hard-edge" shapes which also characterize current passenger car designs – from Audi to Renault. The entertainment and information program built into the seat offers – in addition to video remote control with more than twenty options – a so-called "gamepad." This is a response to the "Gameboy" and its successors, sold by Nintendo many millions of times over.

In outfitting the Airbus A340/600 the "Jochen Müller and Jens Romca Industrial Design" office in 2003 developed new aircraft seats which help eliminate crowding in the aisles and thus facilitate boarding and deplaning – and aircraft cleaning, as well. The dimensions seen here are more modest but the design office is hardly to blame for this. Although the craftsmanship in these seats is excellent, legroom is once again relatively sparse. These seats have, however, not yet been put into service by Lufthansa.

Just as complex as the passenger seats are the so-called "wet zones" or service areas – the galleys and the toilets. In the airframes used for domestic and intra-European flights the galleys are usually located in the crosswalk between two doors. The flight attendants have to prepare meals and beverages on minimized work surfaces, with severely limited storage space. Refreshments are then taken to the passengers using carts or trolleys, depending upon the class of service. These galley walls are almost always made of honeycomb material while the work surfaces, profiles and latches are made of aluminum, which is as light as it is resistant. As a rule, galley walls are lined with containers, drawers and beverage bins stacked one above the other. Their sober functionalism is essentially devoid of any decoration and ornamentation. Their design vernacular is reminiscent of the cool professionalism of commercial kitchens – quite obviously the source of the galley esthetics. Thus the galley is in sharp contrast to the emotionally charged cabin interior with which the passengers have direct contact. In the best case the serving trolley in First Class exudes a certain degree of formal elegance. And then, naturally, there is the bar on the upper deck of intercontinental aircraft such as the Boeing 747, conceptualized at a time when packing in the largest possible number of passengers was not yet the major criterion for efficiency. While the 1959 galley in the Boeing 707, where up to 168 meals could be prepared, still features the typical bubble gum colors of that decade – light yellow, sky blue and pink, the galley on the main deck of a Boeing 747 in 1970 was far larger and laid out with an eye toward efficiency. With temperature sensors between the container compartments and recessed grips, it projects an atmosphere clearly reminiscent of super-rational laboratory conditions. The galley equipment in one embodiment of a Boeing 747 in the early 1970s, however, was adjacent to a generously curved stairway to the upper deck, one which would do honor to any suburban villa and which sometimes provoked lyrical comments.

"Facing stairway to heaven: in the days before fuel crisis and subsequent policies of maximizing passenger count per square meter, the upper deck of the Boeing 747 housed a cocktail lounge while space under the stairs served as a drinks cabinet."[8]

Bordküche einer Lockheed Super Constellation der Lufthansa, 1955–1967 im Einsatz
Galley of a Lockheed Super Constellation operated by Lufthansa, 1955–1967

Priestman Goode: Bordküche, Business Class, Airbus A340-600, ab 2004
Priestman Goode: Business Class galley, Airbus A340-600, as of 2004

großzügig geschwungenen Treppe zum Oberdeck untergebracht, die jedem Einfamilienhaus Ehre machen würde und manchmal geradezu lyrische Kommentare auslöste:

„Treppen zum Himmel: kurz vor der Ölkrise und der damit zusammenhängenden Politik einer Maximierung der Passagieranzahl bot das Oberdeck der Boeing 747 eine Cocktaillounge und unter der großzügigen Treppe eine veritable Zubereitungsküche.“[8]

Der zweite Nasszellenbereich, die Passagiertoiletten, unterliegen zwar den gleichen räumlichen Restriktionen wie die Bordküchen, aber als Räume mit „Publikumskontakt“ muss ihre Anmutungsqualität Sicherheit und Komfort, Hygiene und Freundlichkeit ausstrahlen. Insofern gehört der Entwurf solcher „Lavatories“ zu den komplexesten Designaufgaben im Interieurbereich von Flugzeugen überhaupt. Die vornehm „Erfrischungsräume“ genannten Nasszellen in Flugzeugen mit WC und Waschbecken sind wohl auch erst in den Mittfünfzigerjahren Standard geworden. Denn erst mit der Etablierung von Langstreckenflügen wurden die „natürlichen Bedürfnisse“ unabweisbar. Auch hier wieder ist das primäre Entwurfsproblem der Platzmangel, obwohl Photographien der 1960er Jahre veritable Schminksalons in Flugzeugen zeigen. Aber mit der Kompaktierung der Fluggastsitze wurden auch diese „restrooms“ auf Flächen zwischen 2,5 und 3,5 m reduziert. Und wer hätte nicht schon einmal in einer Personenschlange von sechs bis acht Menschen nach dem Genuss des Flugmenüs und diversen Drinks sich ungeduldig die Beine vertreten? Unvergessen auch jene Damen leicht fortgeschrittenen Alters, die just auf der Mitte eines Interkontinentalfluges ihr Outfit zeitintensiv in diesen „Lavatories“ aufzufrischen beliebten.

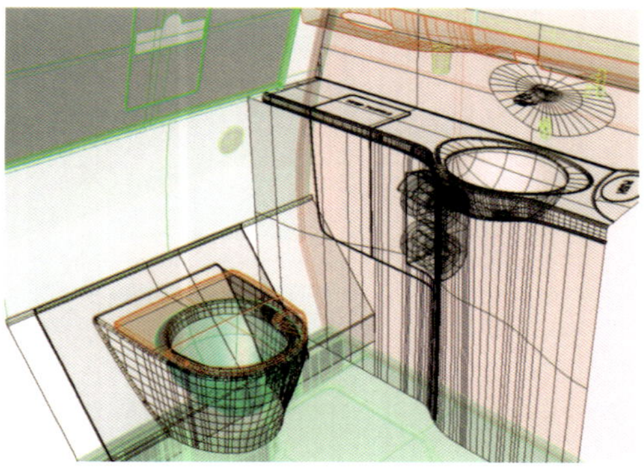

Jochen Müller, Jens Romca: Entwurf Toiletten-Modul, Airbus A340, 2002
Jochen Müller, Jens Romca: Draft for lavatory module, Airbus A340, 2002

Am Beispiel eines aktuellen Entwurfs vom Beginn des Jahres 2004 für Nasszellen im Airbus 300 des Hamburg/Kieler Designbüros Müller Romca mag dies deutlich werden. Es geht um ein humanverträgliches „High-Tech“-Ambiente, dem man weder die unglaubliche Fülle von Randbedingungen noch den Kostendruck ansieht. Jens Romca beschreibt die Aufgabe: „Toilette, Waschraum, Behindertentoilette und separates Urinal sollten entstehen – und das auf ganzen 3,5 Quadratmetern.“[9] Neben einer höchst ökonomischen Raumaufteilung sind weitere Faktoren entwurfsbestimmend: so etwa Griffentfernungen, mit denen selbst Kinder noch zurechtkommen; Waschbecken, die so konstruiert sind, dass sie Spritzwasser auf Armaturenablagen und Fußböden möglichst verhindern; Armaturen, die selbsterklärend ihre Funktion verdeutlichen, wenn der Passagier etwa die Temperatur regelt; Ablagen, die man nicht erst lange suchen muß und Stauraum, der nicht stört, sowie eine Beleuchtung, die nicht blendet. Eine Designfachzeitschrift erläutert:

„Nicht zuletzt muss ein Raumkonzept entstehen, das aus Kostengründen Standardlösungen, aus Ästhetikgründen ein angenehmes Raumgefühl und aus Hygienegründen eine leichte Reinigung zulässt. Und dies alles in einer Formensprache, die auch in den kommenden 30 Jahren noch Akzeptanz finden wird – 30 Jahre, das ist die durchschnittliche Lebenserwartung eines Flugzeugs. Und Lavatories müssen diese Zeit überstehen – sie tauscht man im Regelfall im Laufe eines Flugzeuglebens nicht aus.“[10]

Die Entwürfe des Hamburger Designbüros sind inzwischen realisiert und in die Airbus-Modelle eingebaut. Hinzu kam ein weiteres Problem: Der Raum, der in diesem Flieger den Toiletten zur Verfügung steht, ist unterschiedlich konfiguriert. Denn entweder liegen die Toiletten an der Außenwand des gerundeten Flugzeugrumpfes, was zwangsweise zur Verjüngung der Kabinen im oberen Bereich führen muss, oder sie sind im Unterdeck positioniert, wo sich die Raumbedingungen umkehren, der Fußboden also schmaler, die Raumdecke großzügiger ausfallen muss. Die dritte Version sind Sanitärkabinen, die frei im Raum stehen, also an allen Seiten keine schrägen Wände haben, so dass Decke und Fußboden die gleichen Ausmaße aufweisen. Und da wird dann das

The second “wet zone,” the passenger toilets, is subject to the same spatial restrictions as the galley but as areas for public use they must project a feeling of safety and comfort, hygiene and friendliness. Engineering the toilets is among the most complex design tasks associated with aircraft interiors. The sanitary facilities with a WC and washbasin, discreteely referred to as “lavatories,” did not become standard in aircraft until the mid-1950s. It was only with the advent of long-distance flying that toilets had to be provided. Here again, the primary engineering problem is the shortage of space, even though photographs from the 1960s show veritable cosmetic salons in aircraft. As with the compaction of passenger seating, these restrooms were also reduced to between 2.5 and 3.5 square meters of floor space. And who has not at some time, after enjoying an inflight meal and various drinks, stood in a queue of six to eight people, hopping impatiently from one leg to the other? Unforgettable are those blue-haired ladies who, right in the middle of an intercontinental flight, block the lavatories while devoting careful attention to freshening their make-up – for what seems like hours on end.

All this can be made clear using a current design as an example. The lavatory for the A300 was devised at the beginning of the year 2004 by the Müller and Romca Design Office in Hamburg/Kiel. Here it was necessary to achieve a human but high-tech atmosphere, showing no signs of the many limiting parameters or the cost constraints. Jens Romca describes the task: “Packing a toilet, washroom, handicapped toilet and separate urinal into no more than 3.5 square meters of floor space.”[9] In addition to a highly economical spatial arrangement, further factors have an impact on the design. They include handles and grabs at heights which even small children can reach, washbasins designed to prevent splashing on the countertop and floors, faucets and fittings where temperature regulation is self-explanatory, readily visible areas to accept small personal items, storage space that is neither obvious nor disturbing, and illumination that does not glare or dazzle. One design magazine explained:

“Ultimately it is necessary to devise a space concept which, for cost reasons, permits standardized solutions, while at the same time creating a pleasant spatial feeling in the interest of esthetics and facilitating cleaning and hygiene. And all this has to be done in a design which will find acceptance for the coming thirty years, since that is an airplane’s average ‘life expectancy.’ And lavatories have to survive this entire period. As a rule they are not replaced in the course of an airplane’s service life.”[10]

The concepts worked out by the Hamburg design office have in the meantime been realized and installed in the Airbus models. There was another problem which had to be dealt with, however. The spaces available for toilets in this airframe differ in shape. The toilets at the outside wall of the rounded fuselage will unavoidably have a narrower profile at the top while those located on the lower deck are just the opposite. The room’s cross section is then inverted and thus narrower toward the floor. The third version is the lavatory positioned at the center of the aircraft, having vertical walls, so that the ceiling and floor are identical in size. And this is where the cost problem becomes relevant. Jochen Müller explains: “This doesn’t sound particularly spectacular, but confronts the developers with a huge problem … the numbers produced are too low to justify working out a specific design for each individual cubicle shape.”[11] Thus it is necessary to work with standardized solutions in spite of varying room shapes.

In summary one can state that since the advent of jet aircraft at the end of the 1960s, more and more attention has been devoted to entertainment in designing the interiors, since double-digit hours of flying high over the clouds offer hardly any other diversion.

“Everything from piano bars to a choice of ten ‘radio’ channels with stereo-headphones became the norm. Meanwhile, the cabins of the jets got larger … Not all the

Entwurf Treppe zum Oberdeck, Airbus A380 Jumbo, 2004
Design study for stairs to upper deck, Airbus A380 Jumbo, 2004

Kostenproblem relevant. Jochen Müller erläutert: „Das hört sich unspektakulär an, liefert den Entwicklern aber ein Riesenproblem ... es gibt einfach viel zu geringe Stückzahlen, als dass es sich lohnen würde, für jede Kabinenform eine ganz eigene Konstruktion zu realisieren."[11] Trotz variierender Raumraster also muss hier mit Standardlösungen gearbeitet werden.

Zusammenfassend lässt sich sagen: Seit dem Zeitalter der Jet-Flugzeuge vom Ende der 1960er Jahre an wurde das Interieur immer mehr in den Dienst von Unterhaltungsangeboten gestellt, denn zweistellige Flugstunden hoch über den Wolken bieten ansonsten kaum Ablenkung:

„Alles zwischen einer Pianobar und zehn wählbaren Radiokanälen mit Stereokopfhörern ist inzwischen üblich. Die Kabinen der Jets wurden immer größer ... Zwar hielt nicht jedes erweiterte Raumangebot das, was es versprach, aber vor allem die Kabinenbeleuchtung diente nun nicht mehr nur der optischen Vergrößerung der Kabinen, sondern ebenso dem Erzeugen verschiedenen Atmosphären; nun konnten die Flugbegleiter die Stimmung der Passagiere dadurch beeinflussen, dass sie die Beleuchtung von ‚Schlafzeit' auf ‚Morgengrauen' umstellten ... Um die Mitte der 1970er Jahre wurde der Ruf nach größeren Passagierkapazitäten der Flugzeuge dringender, veranschlagte man doch damals bereits eine Fluggastzahl von 500 Millionen Passagieren pro Jahr. Der ab 1969 in Dienst genommene Jumbo-Jet Boeing 747 war die Antwort auf die ständig wachsende Zahl von Flugpassagieren: mit einer Kabinenbreite von über 6 m, einer Höhe von ca. 2,50 m, einer Lounge im Oberdeck und einer Kapazität von nahezu 500 Fluggästen." In den beiden Gängen zwischen den Sitzen gab es reichlich Raum zum Flanieren und sowohl die Fluggesellschaften wie die Flugzeughersteller nutzten das große Raumangebot für Unterhaltungsprogramme und gesteigerte Bequemlichkeit. Boeing etwa experimentierte mit Kabineneinrichtungen, indem sie zukünftige Passagiere in 1:1-Modelle der 747-Kabinen als Teil einer Produktentwicklungs-Studie einluden.[12]

Immer größere Flugzeuge werden entwickelt. Für den neuen doppelstöckigen Super-Jumbo Airbus A380, im Januar 2005 in Toulouse vorgestellt, wurden Entwicklungskosten von mehr als acht Milliarden Pfund veranschlagt. Aber mit 500 bis 600 Passagieren und einer Reichweite von 15.000 km werden die Kosten pro Passagierkilometer 15 % niedriger sein als etwa bei der Boeing 747. Die Interieurs sind vom Feinsten und erinnern fast schon wieder an die Ozeanriesen der Meere oder an luxuriöse Eisenbahnausstattungen:

„Der neue Super Jumbo wird zwischen den Passagierdecks genug Platz für Cafés, Besprechungs- und Sporträume sowie zusätzliche Toiletten und Schlafkabinen haben. Das Layout der Kabinen ist flexibler als jemals; in den frühen 1970er Jahren wurden die Flugzeugausstattungen alle zehn Jahre überarbeitet, heute alle vier. So haben die Designer des neuen Airbusses verbindliche Vorgaben bezüglich der Einbauten auf ein Minimum reduziert wie z. B. Bordküchen und Toiletten. So können die verschiedenen Fluggesellschaften ihre eigenen Kabinenausstattungen schneller und preisgünstiger installieren. Auch die Gepäckbereiche wurden überarbeitet. Während erster Marktanalysen für den A380 Jumbo war einer der Hauptkritikpunkte von Passagieren das Abhandenkommen von Gepäck bzw. die langen Wartezeiten bei Reklamationen. So entwarfen die Airbus-Designer spezielle Gepäckareale für die First und die Business Class, damit Passagiere mit mehreren Gepäckstücken diese dem Flugbegleiter vor dem Start aushändigen können und nach der Landung zurückbekommen."[13]

So wird die „Klassen"-Gesellschaft auch über den Wolken wohl kaum abgeschafft werden. Geschäftsreisen sind nun einmal etwas anderes als Touristik-Flüge. Insofern spiegeln die unterschiedlichen Komfort-Levels im Passagierflug die unterschiedlichen Bedürf-

increased space was what it seemed, however – lighting played an important part, not only in enlarging the interior space but also in creating different ambiences; the cabin crew could now control the mood of the passengers by changing the lighting from 'night time sleepy' to 'new day dawning' mode ... By the mid-1970s, with scheduled airlines carrying an estimated 500 million passengers per year, the need for real extra space was pressing. The Boeing 747 'Jumbo Jet' (which had been first introduced in 1969) was the answer to the ever-increasing population of air travelers, with its 20-foot-wide cabin, 8-foot-high ceiling, upper-deck lounge and almost 500-strong passenger capacity. There was plenty of room to walk up and down the twin aisles, and airlines and aircraft manufacturers used the space to entertain as well as accommodate. Boeing experimented with interiors by inviting potential passengers to try out mock-ups of 747 cabins as part of a product-development study."[12]

Ever larger aircraft are being developed. Development costs of more than eight billion pounds were budgeted for the new, double-decker Airbus A380 "Super Jumbo," unveiled in Toulouse in January 2005. But with 500 to 600 passengers and a range of 15,000 kilometers, the costs per passenger mile will be about 15% less than those for the Boeing 747, for example. The interiors are exquisite and are once again reminiscent of the great ocean liners or luxurious railway equipments:

"There will be enough space beneath the passenger deck for cafés, meeting rooms, gyms, extra toilets and sleeping cabins. The cabin's layout is to be more flexible; in the early 1970s, cabins were refitted every ten years; now they are redone after four. Airbus' designers have minimized rigid fixtures, like galleys and toilets, so airlines can customize their A380 cabins more quickly and cheaply. The storage areas are also being redesigned. During market research for the A380, one of the main complaints from passengers was lost luggage or the wait at luggage reclaim. Airbus may create special holds in First and Business Class, so passengers with a couple of bags can hand them to the stewards before take-off, then collect them on landing."[13]

And so the "class" society will be perpetuated in the skies. Business travel is, however, something quite different from holiday flights. To this extent the differing comfort levels reflect exactly the differing needs and cost sensitivity of the various customer populations. More space, more comfort and more luxury will, in the future, too, continue to be reserved for the high-price sector because it pulls in more profit for every airline and thus for Lufthansa. On the other hand, democratization in flying is well advanced and is continuing to go forward. For relatively little money one can today fly to New York, Hong Kong, Tokyo or Singapore. But it is exactly this mass business which generates a need for distinction. And aircraft interiors, along with the clearly increasing preference given to frequent travelers, even on the ground, respond to these needs. The red Senator baggage tag, which is still in use today, is perhaps one sign of this distinction.

Entwurf Shopping-Bereich, Airbus A380 Jumbo, 2004
Design study for shopping area, Airbus A380 Jumbo, 2004

**1** Christian Steguweit, „Corporate Identity – Problem oder Lösung für die Fragen der 90er Jahre – Eine Standortbestimmung am Beispiel der Lufthansa", in: Norbert Hammer, *Die stillen Designer – Manager des Designs*, Essen 1994, S. 103–116, S. 114.
**2** Jennifer Coutts Clay, *Jetliner Cabins*, London 2003, S. 178.
**3** Jochen Eisenbrand, „Mehr Beinfreiheit bitte" – Zur Entwicklung des Flugsitzes, in: *Airworld. Design und Architektur für die Flugreise*, Weil am Rhein 2004, S. 124 bis 144, S. 137 f.
**4** Tilman Jens, „Des Kranichs neue Kleider", *Design Report*, Nr. 12, 1994, S. 34 f.
**5** Tilman Jens, „Ein notorischer Vielflieger erlebt das neue Lufthansa Design", *form spezial*, Nr. 2, 1998, S. 66 f.
**6** Tilman Jens, a. a. O., S. 67.
**7** Tilman Jens, a. a. O.
**8** Keith Lovegrove, *Airline. Identity, Design and Culture*, New York 2000, S. 81.
**9** Zit. nach Michael Schweer, „Hightech in Form. Müller/Romca Industriedesign im Porträt", *md*, Nr. 7, 2004, S. 52 ff.
**10** Schweer, a. a. O., S. 53.
**11** Zit. nach Schweer, a. a. O., S. 53 f.
**12** Lovegrove, a. a. O., S. 92, 94.
**13** Lovegrove, a. a. O., S. 105.

Flugzeuginterieur, Tourist Class, Anfang 1980er Jahre
Aircraft interior, Tourist Class, early 1980s

Senator Sleeper, First Class, Boeing 747-200, um 1983
Senator Sleeper, First Class, Boeing 747-200, about 1983

Flugzeugsitze und Bordkino, Boeing 747-200, 1982
Aircraft seats and inflight movie projection, Boeing 747-200, 1982

nisse und Kostenverhalten differenter Bevölkerungsgruppen getreulich wieder. Mehr Raum, mehr Komfort, mehr Luxus wird auch in Zukunft dem Hochpreis-Sektor vorbehalten bleiben, weil er allen Fluggesellschaften, und so auch der Lufthansa, mehr Gewinn bringt. Andererseits ist die Demokratisierung des Fliegens weit fortgeschritten und schreitet weiter fort. Auch für relativ wenig Geld kann man heute nach New York, Hongkong, Tokio oder Singapur fliegen. Aber gerade das Massengeschäft schafft Distinktionsbedürfnisse: Und das Interieur der Flugzeuge ebenso wie die deutlich zunehmende Bevorzugung der „Frequent Travellers" schon am Boden reagiert auf diese Bedürfnisse. Der rote Senator-Kofferanhänger, der bis heute in Gebrauch ist, ist vielleicht ebenso ein Zeichen dieser Distinktion.

Walter Dorwin Teague und Designabteilung der Lufthansa: Kabinengestaltung First Class, Boeing 707-330, 1984
Walter Dorwin Teague and Lufthansa design department: First Class cabin design, Boeing 707-330, 1984

**1** Christian Steguweit, "Corporate Identity – Problem oder Lösung für die Fragen der 90er Jahre – Eine Standortbestimmung am Beispiel der Lufthansa," in Norbert Hammer, *Die stillen Designer – Manager des Designs*, Essen, 1994, pp. 103–116, here p. 114.
**2** Jennifer Coutts Clay, *Jetliner Cabins*, London, 2003, p. 178.
**3** Jochen Eisenbrand "Mehr Eeinfreiheit bitte – Zur Entwicklung des Flugsitzes," in: *Airworld – Design und Architektur für die Flugreise*, Weil am Rhein, 2004, pp. 124–144, here pp. 137 f.
**4** Tilman Jens, "Des Kranichs neue Kleider," *Design Report*, no. 12, 1994, pp. 34 f.
**5** Tilman Jens, "Ein notorischer Vielflieger erlebt das neue Lufthansa Design," in: *form spezial*, no. 2, 1998, pp. 66 f.
**6** Tilman Jens, op. cit., p. 67.
**7** Tilman Jens, op. cit.
**8** Keith Lovegrove, *Airline. Identity, Design and Culture*, New York, 2000, p. 87.
**9** Quotation after Michael Schweer, "Hightech in Form. Müller/Romca Industriedesign im Portät," *md*, no. 7, 2004, pp. 52 ff.
**10** Michael Schweer, op. cit., p. 53.
**11** Quoted after Schweer, op. cit., pp. 53 f.
**12** Keith Lovegrove, op. cit., pp. 92, 94.
**13** Keith Lovegrove, op. cit., p. 105.

Walter Dorwin Teague und Designabteilung der Lufthansa: Studie Sitzbezüge First Class, 1982
Walter Dorwin Teague and Lufthansa design department: Study for First Class seat upholstery material, 1982

Walter Dorwin Teague und Designabteilung der Lufthansa: Studie Sitzbezüge Economy Class, 1982
Walter Dorwin Teague and Lufthansa design department: Study for Economy Class seat upholstery material, 1982

Walter Dorwin Teague und Designabteilung der Lufthansa: Studie Sitzbezüge Economy Class, 1982
Walter Dorwin Teague and Lufthansa design department: Study for Economy Class seat upholstery material, 1982

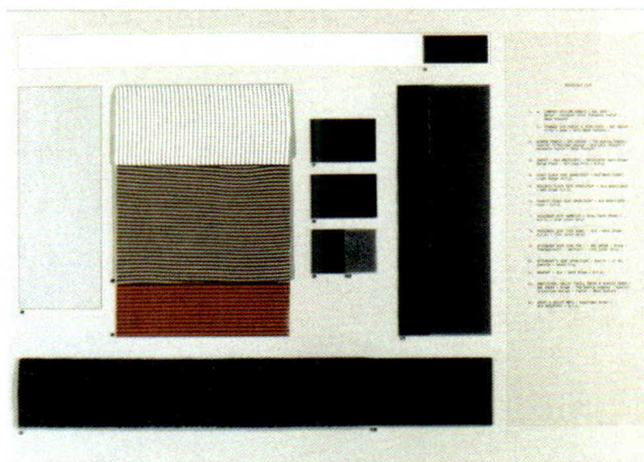

Stoffmuster für die First (hellgrau), die Business (braun) und die Economy Class (rot), 1982, Hersteller: Lantal (ehem. Langenthal)
Fabric samples for First (light gray), Business (brown) and Economy Class (red), 1982, manufacturer: Lantal (formerly Langenthal)

Zintzmeyer & Lux: Flugzeugsitze, First-Class-Kabine, 1990–1996
Zintzmeyer & Lux: Aircraft seats, First Class cabin, 1990–1996

Zintzmeyer & Lux: Flugzeugsitze, Economy-Class-Kabine, 1990–1996
Zintzmeyer & Lux: Aircraft seats, Economy Class cabin, 1990–1996

Zintzmeyer & Lux: Flugzeugsitze, First Class, Schlafposition, 1990–1996
Zintzmeyer & Lux: Aircraft seats, First Class, sleeping position, 1990–1996

Zintzmeyer & Lux: Flugzeugsitze, First Class, Detail, 1990–1996
Zintzmeyer & Lux: Aircraft seats, First Class, detail, 1990–1996

Junkers Ju 52 mit Wellblech-Beplankung, ab ca 1930 – Vorbild für das „Retrodesign"-Konzept von Frog Design
Junkers Ju 52 with corrugated metal skin, as of about 1930 – the model for the "retro-design" concept developed by Frog Design

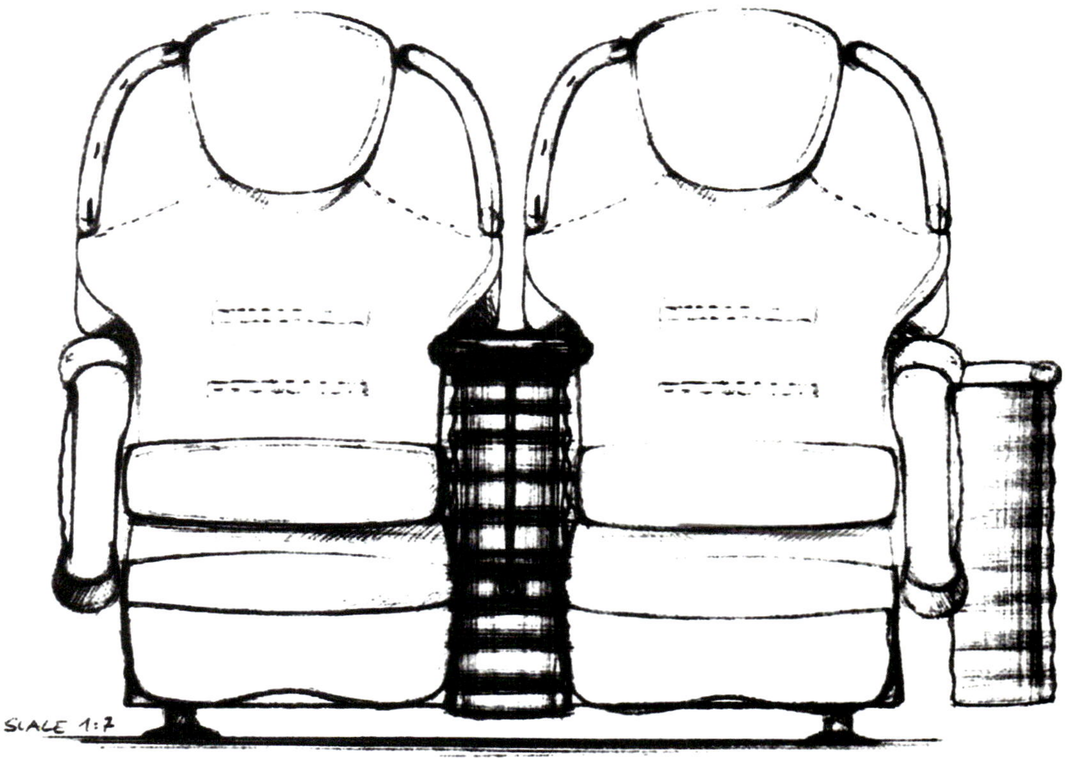

Frog Design: Skizze Flugzeugsitz Business Class, 1996
Frog Design: Sketch for Business Class seats, 1996

Frog Design: Modell Flugzeugsitz First Class, 1996
Frog Design: Draft model for First Class seat, 1996

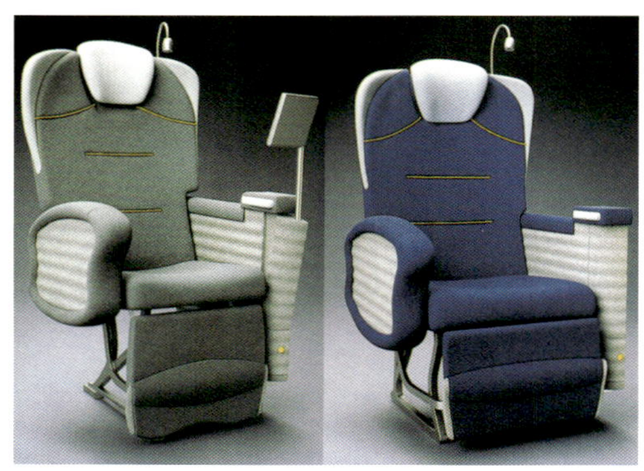

Frog Design: Entwurfsmodelle Flugzeugsitz Business Class, 1996
Frog Design: Draft models for Business Class seats, 1996

Frog Design: Flugzeugsitze, First Class, ab 1998
Frog Design: Aircraft seats, First Class, as of 1998

Frog Design: Flugzeugsitze, Business Class, ab 1998
Frog Design: Aircraft seats, Business Class, as of 1998

Priestman Goode: Flugzeugsitze Business Class mit integriertem Inseat-Video, ab 2004
Priestman Goode: Aircraft seats, Business Class, with integrated inseat video, as of 2004

Priestman Goode: Master-Control-Einheit für Inflight Entertainment und Sitzsteuerung, Business Class, ab 2004
Priestman Goode: Master control unit for inflight entertainment and seat controls, Business Class, as of 2004

Priestman Goode: Flugzeugsitze Business Class, "Private Bed"-Position, ab 2004
Priestman Goode: Aircraft seats, Business Class, in the "private bed" position, as of 2004

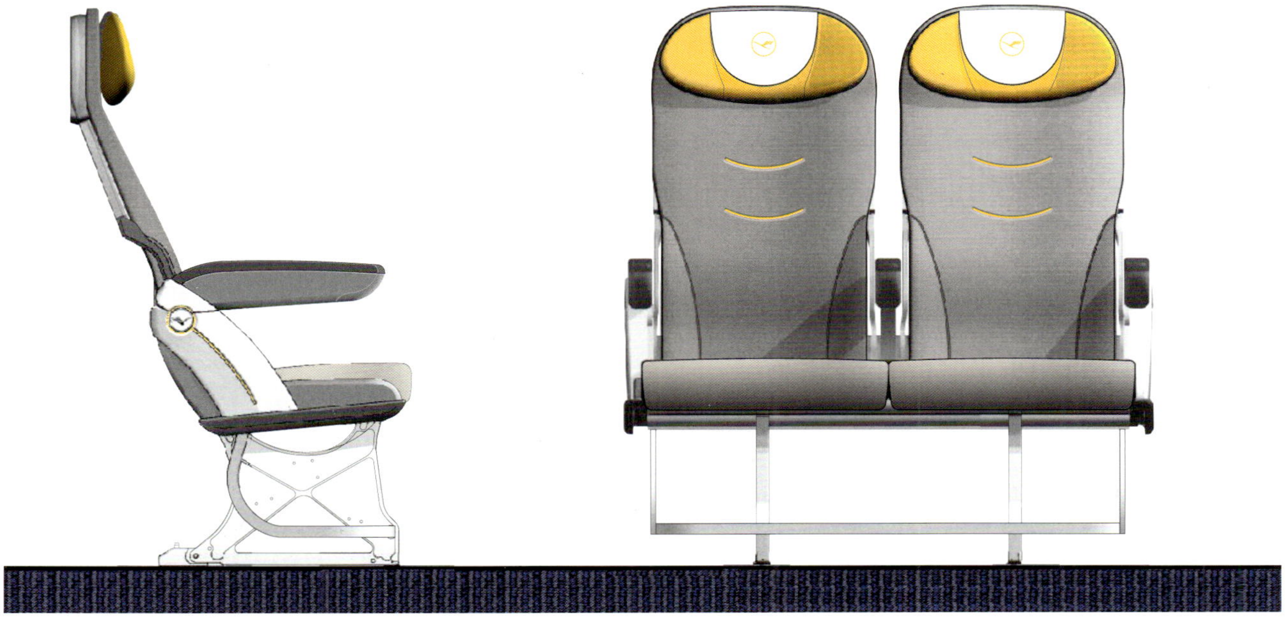

Priestman Goode: Entwurf Flugzeugsitz Economy Class, 2004
Priestman Goode: Draft for aircraft seat, Economy Class seat, 2004

Priestman Goode: Kabineninterieur Economy Class Airbus A340-600, ab 2004
Priestman Goode: Cabin interior Economy Class, Airbus A340-600, as of 2004

# „Das Auge isst mit"

Sobald die zivile Luftfahrt in der zweiten Hälfte der 1920er Jahre geschlossene Fluggastkabinen hatte, entstand bei den Passagieren das Bedürfnis nach Zerstreuung und Komfort an Bord. Neben Zeitungen und Zeitschriften sowie mitgeführten eigenen Büchern, neben den dann später entstehenden Musik- und Filmangeboten in den Maschinen besteht dieser Komfort nicht zuletzt aus dem Angebot an Essen und Trinken in angenehmer Ausstattung und qualitätvollen Table Tops. Das gastronomische Ambiente definiert das Erscheinungsbild einer Fluggesellschaft entscheidend mit. Keith Lovegrove bemerkt:

„Das Corporate Design war und ist ein wichtiges Mittel und so wie die Fluggesellschaften ihre jeweils eigene Flugzeugaußenkennzeichnung entwickelt haben, so haben sie dies auch in den Flugzeugen getan. So bieten Agenturen einen speziellen Entwurfsservice für Bordbestecke oder Bordgeschirre an. Dabei ist eine Entwicklung zu beobachten, die von dem rektangulär verkettbaren Essschalen unter dem Motto ‚möglichst viele pro Tablett' sich zugunsten einer mehr organischen Formgebung hin entwickelt, die sowohl von der generellen Corporate Identity einer Fluggesellschaft als auch von nationalen Identitäten inspiriert sein kann. Gleichzeitig wird die Positionierung des Logos auf dem Bordgeschirr aus ästhetischen Gründen zunehmend zurückhaltender, – nicht zuletzt aber auch deshalb, weil auffällig gekennzeichnete Bordmittel gerne als Trophäe mitgenommen wird."[1]

Vor allem für die Menüs weisen die Passagiersitze ausklappbare Tablare auf, die die auf Tabletts vom Bordpersonal servierten Speisen sicher über dem Schoß des Fluggastes fixieren helfen. Das Handling allerdings dieser gastronomischen Angebote in Flugzeugen unterliegt spezifischen Anforderungen und Einschränkungen (vgl. auch das Kapitel Bordausstattung). Die räumlich beengte Anlieferung durch die Gänge der Maschinen an die einzelnen Passagiersitze erforderte bis weit in die 1960er Jahre bescheiden dimensionierte Servierwagen, seitdem vertikal gestaltete Roll-Caddys, in denen die fertig konfektionierten Speisen, von externen Catering-Unternehmen geliefert, in Schubladen-Tablaren lagern. Jedenfalls ist dies seit den 1960er Jahren so: Bis dahin nämlich gab es noch Köche an Bord, die individuell die Speisen in der Maschine zubereiteten. Diese Rollwagen, die tendenziell die Anmutung und den etwas technoiden Charme eines mobilen Werkzeugwagens haben, krönen die Batterien von Getränken, eine veritable Mischung aus diversen Bieren und Weinen, Wässern und Fruchtsäften sowie Spirituosen. Hinzu kommen Kaffee- und Teekannen, aus denen diese Heißgetränke frisch gebrüht serviert werden. Diese Rollwagen – Logistikbehälter auf komprimiertestem Raum – müssen anderen ergonomischen Anforderungen genügen als das Bordgeschirr selbst. Bei diesem werden zusätzlich die Ästhetik und Gestaltung wichtig, weil es unmittelbar mit den Fluggästen in Kontakt kommt und von ihnen individuell benutzt wird. Entsprechend variantenreich sind diese Bordgeschirre bei verschiedenen Fluggesellschaften und je nach Kulturkreis, Formvorlieben der Entwerfer und der sich wandelnden Corporate Identity der Fluggesellschaften gestaltet. Bordgeschirre sind neben den Passagiersitzen jene Produktgruppe, mit der am deutlichsten der Komfort einer Flugreise verbunden wird: ergänzt natürlich durch die Qualität der dargereichten Speisen. In einem Essay über „Geschirr in der Luft" bemerkt die Autorin schon 1967 hellsichtig:

„Auch der Fluggast, der gern unentwegt hinunterschaut auf Wolken, Wasser und Inseln, wendet sich meist seinem Tablett zu. Das Essen im Flugzeug ist für einige

Seite 114: Wolf Karnagel: Bordbesteck, ab 1986
Page 114: Wolf Karnagel: Cutlery, as of 1986

# "Appetite begins in the eye of the beholder"

Once passenger cabins had been enclosed in civil aviation – in the latter half of the 1920s – passengers soon discovered the need for diversion and additional comfort while on board. The first such diversions were in the form of newspapers and magazines – and books which passengers might have brought with them. These were joined much later by inflight audio and then video. Major components in making passengers feel comfortable while airborne are foods and beverages offered using an appealing table service and quality place settings. Everything associated with food service makes a decisive impact on defining an airline's corporate image. Keith Lovegrove commented:

"Corporate image was and is a very important tool and just as airlines have developed their own unique livery outside the aircraft, they have also desired their own unique style inside. Identity consultancies offer a product design service to create distinctive cutlery and crockery shapes. There is a move away from square plastic, get-as-many-dishes-on-a-tray culture to more organic, oval shapes that are inspired by elements of either the corporate identity or national heritage. The display of the logo on crockery is becoming more subliminal for aesthetic reasons but also because there is evidence that an airline-branded item has more trophy-status."[1]

The passenger seats are equipped with fold-down tray tables, primarily for use with the complete meals. These tray tables help to fix in place, securely over the passenger's lap, the food which flight attendants serve on specially designed trays. Handling the range of foods and beverages offered in the air is, however, subject to specific requirements and restrictions (see also the discussion in the chapter on cabin interiors). Delivering meals to passengers in limited space, down the aisles of the aircraft to the individual rows of seats, required compact serving trolleys until well into the 1960s. They were succeeded by upright wheeled carts in which finished meals, delivered by outside catering companies, are stored on trays resting on rails inside the cart. This represented a major change in the food service concept. Up to that time, the 1960s, cooks had prepared individual meals on board. These carts, which tend to have the appearance and technical charm of a wheeled toolbox, are crowned with a veritable battery of beverages – a broad selection of beers and wines, waters and fruit juices and spirits, too. Also to be found atop the carts are containers for freshly brewed coffee and tea. These wheeled carts – logistics units engineered to occupy minimum space – have to satisfy different ergonomic requirements than does the dinnerware itself. Esthetics and appearance are of major importance in the latter, as passengers handle and use them directly and individually. Accordingly, these utensils display a remarkable variety, depending on the airline itself and the culture of its home country, the designers' preferences in regard to shapes, and the carriers' ever-changing corporate identity. Second only to the passenger seats, these inflight utensils are the items most closely associated with the degree of comfort offered on an air journey, augmented, of course, by the quality of the foods being offered. In an essay on "Utensils in the air" the author noted almost prophetically in 1967:

"Even passengers who enjoy watching the clouds, water and islands for hours on end will usually turn their attention to the tray. inflight meals are for some airlines a question of prestige but they are also a form of occupational therapy; passengers who are eating remain quietly seated and do not wander around the cabin. Only a few will

Senator-Service, First Class, Boeing 707, um 1965
Senator service, First Class, Boeing 707, about 1965

Senator-Service, First Class, Boeing 707, Bordsteward, um 1965
Senator service, First Class, Boeing 707, steward, about 1965

Fluggesellschaften eine Prestigefrage, es ist aber auch eine Beschäftigungstherapie: Der essende Passagier sitzt ruhig und läuft nicht umher; nur wenige schlagen aus, was ihnen, stets lächelnd, serviert wird. Luftfahrtgeschirr und Besteck sind Visitenkarten der Gesellschaft und des Herkunftslandes. Man umwirbt auch damit die Fluggäste. Das Essen soll ihnen eine angenehme Beschäftigung sein“.[2]

Aber jenseits aller nobilitierenden Romantik des Gourmetgenusses über den Wolken unterliegt das Speisen im Flugzeug den logistischen Bedingungen einer Betriebskantine. Die Anlieferung der Nahrung und der Getränke an die Maschinen, ihre Unterbringung in den räumlich begrenzten Bordküchen und Caddys, die Stapelbarkeit der Tabletts (auch gefüllt) und der einzelnen Geschirrteile, die Rücknahme der benutzten Geschirre und ihre Reinigung: All dies ist nur in großserienhaftem Maßstab machbar, abzuwickeln und wirtschaftlich zu kalkulieren. Auch die Auswahl und Zusammenstellung der Speisen, die Größe der Portionen, die Konditionen der Lebensmittellieferungen, die verschiedenen Komplexitätsgrade der Zubereitung und Portionenkonfektionierung am Boden sind hart kalkulierte ökonomische Faktoren. Obwohl die zehn bis 15 Euro pro 4-Gänge-Menü inklusiv Getränken so gravierend nicht erscheinen, beeinflussen sie den Flugpreis schon. Andererseits ist der Imagefaktor eines ansprechend dargebotenen und luxuriös-frischen Flugmenüs sicherlich höher zu bewerten als die betriebwirtschaftliche Minimax-Strategie verantwortlicher Betriebsleiter.

Doch bleiben wir bei den Gefäßen und ihrer Formgebung. Neben allen gestalterischen Unterschieden der verschiedenen Modelle unterliegen die Speisebehälter und Bestecke flugreisespezifischen Anforderungen. Der Neigungswinkel nämlich zwischen Passagiersitz und dem über dem Schoß des Fluggastes platzierten Tablett ist naturgemäß eng und steil und jeder Passagier hat tatsächlich nur den Platz der Tablettgröße zur Speiseaufnahme und zur Handhabung des Bestecks zur Verfügung. Dementsprechend müssen die Bestecke gedrungener sein als jene für die normale Gastronomie oder den privaten Haushalt. Und auf dem Tablett gibt es spätestens seit den 1960er Jahren so gut wie immer eine modulare, in der Regel rektanguläre Anordnung einzelner Schalen, die die verschiedenen Gänge eines Flugmenüs aufnehmen. Nicht zuletzt geht es – wie überall beim Fliegen – um ein möglichst geringes Gewicht. Deshalb auch haben vor allem in der Economy Class Kunststoffe die Porzellangeschirre ersetzt, die in der Regel inzwischen nur noch in der Business- und in der First Class eingesetzt werden. Alle Speisebehälter weisen minimalisierte Einzelportionen auf und werden bereits am Boden von der LSG (Lufthansa Service Gesellschaft) fertig verpackt und an die Maschinen angeliefert. Die Speiseschalen sind in der Regel verletzungsresistent abgerundet, aber doch so spezifisch gestaltet, dass ihre Form als „typische“ Form der jeweiligen Fluggesellschaft im Gedächtnis bleibt. Die Ober- bzw. Unterflächen der Schalen und Tabletts müssen rutschfest ausgelegt sein, schon wegen nie auszuschließender Turbulenzen während eines Fluges. Die separate Verpackung jeder Komponente des Flugmenüs hat nicht nur hygienische und logistische, sondern auch Auswahlgründe. Dabei halten sich inzwischen alle Fluggesellschaften zugute, dass auf Grund der Einschweißung die Speisen ansprechend und frisch auf den Tisch bzw. das Tablar kommen. Gewürze werden in minimalisierten Tütchen beigegeben, ebenso wie Zucker, Butter, Kondensmilch oder Papier- bzw. Stoffservietten, je nach Flugklasse. Nicht zu vergessen die Zahnstocher und die Erfrischungstüchlein. Bei Long-Distance-Flügen kommen oft heiß-feuchte Handtücher zur Gesichtserfrischung vor der Landung hinzu, dargeboten von den Stewardessen mit aseptisch wirkenden Holzzangen: nach langem Flug also das Angebot einer Wellness-Neugeburt. Und nebenbei: Rauchen im Flieger ist seit Jahren nicht mehr opportun, obwohl die Sitzlehnen immer noch kleine Aschen-

refuse the meals, always served with a smile. Airline crockery and cutlery are the calling cards for the firm and its home country. They are used to attract passengers. Eating is to be a pleasant pastime for them."[2]

But beyond all the grandiloquent romanticizing about gourmet delights above the clouds, dining on board is subject to logistical constraints similar to those found in a company cafeteria. Delivering foods and beverages to the aircraft, stowing them in the tight spaces afforded by the galleys and carts, stackability of the trays (even when full) and the individual meal dishes, collecting and cleaning used dinnerware – all this can be made feasible and economical only at virtually industrial scale. The selection and combination of the individual foods, the portion sizes, commercial terms for food deliveries, and the various degrees of complexity in preparing and making up individual portions on the ground are all profitability factors which have to be carefully calculated. Although the cost, at ten to fifteen euros per four-course meal, including beverages, would not seem to be critical, they do in fact have an influence on the ticket price. On the other hand, the image factor of an appealingly presented, lusciously fresh airborne meal is certainly to be assigned greater weight than the minimize-maximize business strategy applied by the operations managers in charge.

But let's take a closer look at the meal dishes and their shapes. In addition to all the differences in design among the various series, crockery and cutlery have to satisfy certain specific requirements for airborne use. The angle between the passenger's own seat and the tray table hovering above his or her lap is by nature narrow and steep. In actual fact, the passenger has to wield cutlery exclusively in the space defined by the tray itself. Accordingly, cutlery will have to be shorter than that used in restaurants or at home. Ever since the 1960s, at the latest, the serving tray has almost always featured a modular – and usually rectilinear – arrangement of the individual meal dishes which hold the various courses of an airline meal. Ultimately it is also a matter of achieving the lowest possible weight, as is true everywhere in aviation. That is why, in Tourist Class above all, plastic has replaced china, which today will normally be found only in Business and First Class. All the foods are individually portioned and meals are assembled and packed on the ground by the LSG (Lufthansa Service Gesellschaft) for delivery to the aircraft. The individual meal dishes are generally rounded to prevent injury but are still so distinctively designed that their shape is remembered as the "typical" form for the particular airline. The mating surfaces of the dishes and serving trays have to be slip-resistant to help counteract the effects of turbulence during the flight, something which must always be anticipated. Individual packaging of each item in the inflight meal is not only for hygienic and logistical reasons; this also helps in making up special meals. In the meantime all the airlines point with pride to the fact that – due to individual packaging – the foods placed on the tray or table are both appealing and fresh. Salt and pepper are served in tiny paper packets and are accompanied by pre-portioned sugar, butter, condensed milk and dinner napkins in paper or cloth, depending upon the class of service. Nor should we forget the toothpicks and the astringent towelettes. On long-distance flights hot and steamy terry towels are often distributed for freshening up prior to landing. They are handed out by the flight attendants using metal tongs to lend an aseptic touch. After a long flight this offers something of a rebirth for weary travelers. And just in passing: Smoking in airplanes has been frowned upon for years now, but the armrests still feature miniature ashtrays which nowadays might be used only for chewing gum. The flight attendants, by the way, definitely have the last word inside the aircraft. They can most assuredly refuse rowdy passengers, after the fifth whisky, the sixth one.

Don Wollance: Lufthansa-Besteck, 1950 (Re-Edition 2004, limitierte Auflage), Hersteller: Pott
Don Wollance: Lufthansa cutlery, 1950 (re-edition 2004, limited edition), manufacturer: Pott

Let us turn our attention once again to the tableware. It goes without saying that cutlery must not be suitable for use as a weapon. Consequently sharp-pointed steak knives will no longer be found on any airline. It is just as understandable that individual items may never be so heavy that passengers could be injured in the event of turbulence.

Keeping all these limitations in mind, we now want to review the Lufthansa place settings used through the decades. Menus and advertisements have in some cases deliberately been mentioned together with the various meal service products since the products and the graphic designs are closely interrelated. The design questions specific to the graphics are examined in greater depth in the chapter on corporate design and presented in chronological order.

Even prior to the famous Wilhelm Wagenfeld place setting, American designer Don Wollance devised in 1950 the four-piece "Model 21" cutlery service in the "rounded arc" style typical of the era and which was used by Lufthansa from 1955 onward. Quite noticeable here is the flowing, asymmetrical knife handle and the knife blade which ends in a semicircle. This cutlery, with its almost organic flow, appears to be quite Scandinavian on the formal level, similar to that designed by Alvar Aalto for SAS. It was reissued in 2004, on the one-hundredth anniversary of the original manufacturer, the Pott company in Solingen, Germany, as a limited edition of 250 sets in stainless steel. Susanne Prinz carefully documented the creation and design process for this first in-flight cutlery service crafted especially for Lufthansa:

"Through the mediation of American cutlery wholesaler Lauffer, who was presumably interested in the cutlery being from an American design studio, Carl Pott made contact with Don Wollance in 1950. The result of the collaboration between these two men was cutlery design No. 2721. In a version devised by Carl Pott it became the first cutlery to be used by Lufthansa. In many respects it is a product typical of its age. Because, for cost reasons, there was never any personal contact between the German manufacturer and the American designer, all the steps needed to work out the final form had to be clarified by mail. A brochure accompanying the cutlery, written by the American distributor, described in detail the reasons for the shape of each individual item. Wollance's extensive research on eating habits, using the cutlery services customary in Europe and America, resulted in the following design characteristics:

"1. Spoon and fork handles to have a concave upper and convex lower surface in order to ensure safe use of the utensils and a pleasant feeling in the hand. 2. Handles curved so that their profiles follow the shape of the fingers and the palm of the hand. 3. Increased thickness at the necks of the handles to keep the spoons and forks from bending when used to mash foods. 4. Narrowing the convex shape of the spoon and fork handles toward the neck so as not to interfere with the change in the hand's grasp as the utensil is moved from the plate to the mouth. 5. Spoon-like depression at the base of the shortened tines, making it easier to deal with foods such as peas. 6. Trowel-like bowl of the spoon with a longitudinal axis at right angles to the handle. 7. Truncated knife blade at a distinct angle to the handle, in order to make cutting as effortless as possible. 8. Knife handle in a shape derived from the traditional pistol grip form which, nonetheless, permits manufacture as an economical single-piece knife. The first models were then made up in aluminum, incorporating the findings developed during this scientific analysis. Aluminum sheet was chosen because it is easy to cut and shape. The following generations of mock-ups were made of wood and the final proposals, which were then sent to Germany, were made of pewter, which very closely resembles stainless steel in its appearance. The craftsmen at the Pott company were entrusted with

Senator-Service, First Class, mit Chef-Steward, 1958
Senator service, First Class, with chef steward, 1958

Senator-Service, First Class, 1958
Senator service, First Class, 1958

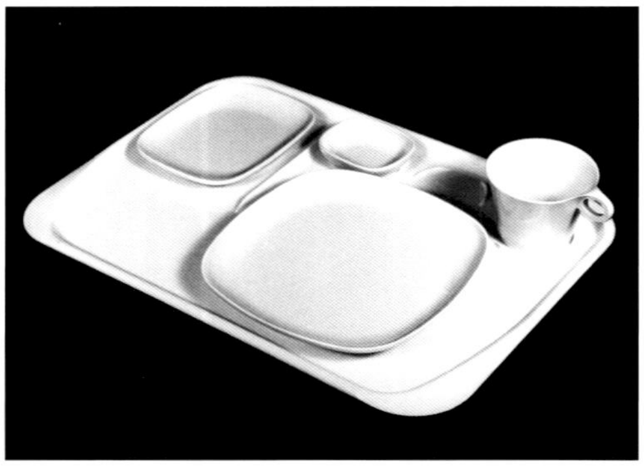

Wilhelm Wagenfeld: Bordgeschirr aus Melamin, Tablett aus gelbem Nylon, 1954/55, Hersteller: Buchsteiner
Wilhelm Wagenfeld: Airline dinnerware made of Melamine, serving tray in yellow nylon, 1954/55, manufacturer: Buchsteiner

becher aufweisen, die inzwischen allerdings nur noch für Kaugummis genutzt werden dürften. Im Übrigen haben die Stewardessen und Flugbegleiter in der Maschine gewissermaßen „Hausrecht". Sie können härteren Naturen nach dem fünften Whisky den sechsten oder siebten durchaus verweigern.

Zurück zum Bordgeschirr: Selbstverständlich darf es sich nicht als Waffe eignen und insofern sind scharf-spitze Steakmesser bei keiner Fluggesellschaft auf der Agenda. Ebenso nachvollziehbar ist, dass die Geschirrteile nicht so schwer sein dürfen, dass sie bei eventuellen Turbulenzen Passagiere verletzen könnten.

All diese Beschränkungen in Rechnung gestellt, schließt sich nun ein Überblick über das Bordgeschirr der Lufthansa quer durch die Jahrzehnte an. Bewusst sind den wechselnden Table-Top-Produkten ab und zu auch Speisekarten und Advertisements zugeordnet, weil sich hier Produkt- und Grafikdesign eng verzahnen. Die grafikspezifischen Entwurfsfragen wurden bereits im Kapitel über Corporate Design vertiefend und in der historischen Abfolge dargestellt.

Noch vor dem dann berühmten Bordgeschirr von Wilhelm Wagenfeld entwarf 1950 der Amerikaner Don Wollance ein vierteiliges Besteck „Modell 21" im typischen „Rundbogenstil" der Zeit, welches die Lufthansa 1955 übernahm. Auffällig ist der asymmetrisch geschwungene Messergriff ebenso wie die halbkreisförmig endende Messerschneide. Dieses Besteck wirkt in seinem organoiden Schwung formal durchaus skandinavisch, wie von Alvar Aalto für die SAS entworfen. Es wurde 2004 zum Jubiläum des 100-jährigen Bestehens des damaligen Herstellers, der Solinger Firma Pott, als limitierte Edition von 250 Stück in Edelstahl neu aufgelegt.

Susanne Prinz hat den Entstehungs- und Entwurfsprozess dieses ersten Bordbestecks der Deutschen Lufthansa akribisch beschrieben: „Durch die Vermittlung des amerikanischen Tafelgerätegroßhändlers Lauffer, dem vermutlich an einem Besteck gelegen war, das von einem amerikanischen Designer stammte, bekommt Carl Pott 1950 Kontakt mit Don Wollance. Das Resultat der Zusammenarbeit dieser beiden Männer ist das Besteck Nr. 2721, das in einer Variante von Carl Pott auch das erste Bordbesteck der Deutschen Lufthansa wurde. Es ist in vieler Hinsicht ein typisches Produkt seiner Zeit. Weil aus Kostengründen niemals ein persönlicher Kontakt zwischen dem deutschen Hersteller und dem amerikanischen Entwerfer zustande kam, war es nötig, alle zur Findung der endgültigen Form notwendigen Schritte in schriftlicher Form zu fixieren. In einem dem Besteck beigegebenen Begleitheft des amerikanischen Vertreibers wurde dem Kunden die Form jedes einzelnen Teils genau begründet. Die umfassenden Recherchen von Wollance über die menschliche Nahrungsaufnahme unter Zuhilfenahme eines in Europa und Amerika üblichen mehrteiligen Bestecks führten zu folgenden Entwurfskennzeichen:

1. Konkave Vorder- und konvexe Rückseite der Stiele von Löffel und Gabel, um eine sichere und für die Hand angenehme Benutzung der Geräte zu gewährleisten. 2. Eine im Profil der Form von Finger und Handfläche folgende Kurvatur der Griffe. 3. Betonte Dicke des Stielhalses, um ein Verbiegen von Löffel und Gabel zu vermeiden, wenn diese zum Zerkleinern von Lebensmitteln benutzt werden. 4. Verengung der konvexen Form der Löffel- und Gabelgriffe zum Hals hin, um den wechselnden Zugriff der Hand, wenn sie sich vom Teller zum Mund bewegt, nicht zu stören. 5. Löffelartige Vertiefung des Gabelschiffs und verkürzte Zinken, damit zum Beispiel auch Erbsen besser aufgenommen werden können. 6. Kellenartige Löffellaffe mit einer im rechten Winkel zum Stil stehenden Längsachse. 7. Verkürzte Messerschneide in einem betonten Winkel zum Griff, um einen möglichst mühelosen Schnitt zu erlauben. 8. Messergriff in einer aus dem traditionellen Pistolengriff entwickelten Form, die dennoch die Herstellung eines

subsequent transformation into mass-production items. The manufacturing process had already been integrated into the shape as a part of the design criteria. Like many of his contemporaries, Wollance wanted to make the creation process for a utensil transparent and did so by selecting a shape which makes the stamping and pressing operations immediately apparent. He was especially successful in this effort with the knives. Their distinctive contour seems to have been cut out of the material and that is precisely the case."[3]

The first of Germany's renowned product designers to be selected to develop crockery and cutlery for Lufthansa was then – as previously mentioned – Wilhelm Wagenfeld. Ever since the Bauhaus era he had been an established designer of china, glass and metalware for the well-laid table. In the 1950s and 1960s his designs dominated the line of products manufactured by the Württembergische Metallwarenfabrik (WMF). His airline utensils, crafted in 1954/55, were a trail-blazing design achievement. Discussed in contemporary trade journals, this series attracted international attention and found many imitators. Responding to the dining culture and serving esthetics of the 1950s Wagenfeld concentrated entirely upon large, shallow dinner plates which could accept generous portions. Contemporary photographs show quite clearly these "mountains of food" prepared on board. In formal terms, these dinner plates are rounded squares – a shape which is just as ergonomically correct as it is logistically sensible since several items of dishware of this shape, in differing sizes, could be arranged at close spacing on the serving tray. Despite all the differences in detail, this "rounded square" continued in the following decades to be a basic figure in design thinking for Lufthansa place settings. The material for the crockery manufactured by the Buchsteiner company was blue-gray Melamine; the serving tray was manufactured from yellow nylon. Just one year after Wagenfeld's dinnerware was introduced, however, a change was made to Luran since the Melamine used originally had proved to be too heavy. Thus the traditional Lufthansa corporate colors, blue and yellow, were used with striking clarity here but were then toned down in the more self-assured and elegant designs of later decades. The salt and pepper shakers matching the Wagenfeld place settings were, by the way, officially designated as souvenirs for the passengers – and they enthusiastically accepted the invitation to pocket them. The great disparity between the design of these place settings and tastes prevailing at that time is demonstrated in a comparison with menus from that period, featuring floral still lives à la Dürer in the background and announcing in all-caps, as though chiseled in marble, "Mid-Morning Snacks," "Appetizing Tidbits" and "Mocha" coffee as the crowning touch. In 1958 Holsten brand beer was even served on draft, tapped from wooden kegs on board. Until well into the 1960s there was, in addition, a stand-up bar with a bartender in the First Class section of the Boeing 707. This was an attempt to recreate the flair and subtly encoded erotic signals found in the "nightlife" celebrated at certain establishments in the major cities. Of course smoking, the bar itself and the occasional flirt there during the flight were among those creature comforts which had not yet been questioned. On intercontinental flights to the Far East there were – at least as contemporary Lufthansa advertising photographs suggested – Asian stewardesses on board even in the 1960s. Seen from today's viewpoint, of course, with almost half a century of hindsight, the patina of such photos is as seductive as it is nostalgic. We suspect, and in fact know, that the era of grandeur in the aviation experience, ostentatiously put on show here, irrevocably became a thing of the past with the advent of mass tourism. Today alert young managers sit hunched over their laptops, hard at work, and are more likely to order fruit juice than a cocktail.

Hans „Nick" Roericht (Entwicklungsgruppe 5 der HfG Ulm): Tablett aus Luran, 1968, Material: BASF
Hans "Nick" Roericht (Development Group 5 at the Ulm Design College): Serving tray made of Luran, 1968, material: BASF

Hans „Nick" Roericht (mit Michiko Asaka und Franco Clivio): Prototypen Bordgeschirr, 1961–1975
Hans "Nick" Roericht (with Michiko Asaka and Franco Clivio): Prototypes for airline dinnerware, 1961–1975

Hans Theo Baumann: Bordgeschirr First Class, 1963, Hersteller: Rosenthal
Hans Theo Baumann: First Class dinnerware, 1963, manufacturer: Rosenthal

Hans „Nick“ Roericht: Bordgeschirr Touristenklasse, ab 1967
Hans "Nick" Roericht: Tourist Class dinnerware, since 1967

Hans „Nick“ Roericht: Bordgeschirr in der Ausstellung „Gut gestaltetes Gebrauchsgut“, Deutscher Pavillon, „Expo 67“, Montreal
Hans "Nick" Roericht: Airline dinnerware at the "Fine design in utility items" exhibition in the German pavilion at the "Expo 67" in Montreal

preisgünstigeren Monoblockmessers erlaubt. Nach diesen auf wissenschaftlichem Weg gefundenen Ergebnissen entstanden die ersten Aluminiummodelle. Aluminiumblech wurde deshalb gewählt, weil es leicht auszuschneiden und zu verformen ist. Die nachfolgende Modellgeneration wurde aus Holz und die endgültigen Vorlagen, die nach Deutschland geschickt wurden, aus einer Zinn-Blei-Legierung gefertigt, die im Farbton dem Edelstahl recht nahe kam. Die anschließende Umsetzung in ein Serienprodukt oblag den Handwerkern der Firma Pott. Der Herstellungsprozess war bereits als ein Entwurfskriterium in die Form eingegangen. Wie viele seiner Zeitgenossen wollte Wollance den Entstehungsprozess eines Besteckes transparent machen, indem er eine Form wählte, die das Stanzen und Pressen unmittelbar vor Augen führt. Besonders die Messer sind ihm in dieser Hinsicht gelungen. Ihre ungewöhnliche Kontur wirkt wie ausgeschnitten, was sie in der Tat ja auch ist.“[3]

Der erste renommierte deutsche Produktgestalter, der für die Lufthansa Bordgeschirre entwickelte, war dann – wie erwähnt – Wilhelm Wagenfeld, ein seit den Zeiten des Bauhauses etablierter Gestalter für Porzellan-, Glas- und Metallwaren für den gedeckten Tisch, der in den 1950er und 1960er Jahren mit seinen Entwürfen etwa auch die Produktpalette der Württembergischen Metallwarenfabrik (WMF) dominierte. Sein Bordgeschirr, entworfen 1954/55, war ein wegweisender gestalterischer Entwurf. In zeitgenössischen Fachpublikationen besprochen, errang diese Serie internationale Aufmerksamkeit und fand viele Nachahmer. Wagenfeld konzentrierte sich entsprechend der Esskultur und Servierästhetik der 1950er Jahre ganz auf große, flache Speiseteller, die üppige Portionen aufnehmen können. Zeitgenössische Aufnahmen zeigen diese an Bord zubereiteten „Speiseberge“ sehr deutlich. Formal sind die Speiseteller durch gerundete Quadrate charakterisiert; eine Form, die ebenso ergonomisch wie logistisch ist, da mehrere solcherart konturierte Gefäße in unterschiedlicher Größe platzsparend auf dem Tablett dicht angeordnet werden können. Dieses „Rundquadrat“ bleibt dann auch weiterhin in den folgenden Jahrzehnten bei allen Unterschieden im Detail eine Grundfigur des entwerferischen Denkens für das Bordgeschirr der Lufthansa. Das Material des von dem Unternehmen Buchsteiner produzierten Geschirrs war blaugraues Melamin, das Tablett wurde aus gelbem Nylon gefertigt. Bereits ein Jahr nach seiner Einführung allerdings wurde Wagenfelds Geschirr aus Luran produziert, da sich Melamin als zu schwer erwiesen hatte. Die traditionellen CI-Farben also der Lufthansa, Blau und Gelb, eingesetzt mit einer plakativen Deutlichkeit, die bei Entwürfen späterer Jahrzehnte selbstbewusst-elegant zurückgenommen wurde. Die Salz- und Pfefferstreuer des Wagenfeld-Geschirrs übrigens waren offiziell als Souvenir für die Passagiere gedacht und diese machten auch reichlich Gebrauch davon. Wie groß der gestalterische Abstand dieses Geschirrs zum vorherrschenden Zeitgeist war, zeigen die zeitgleichen Menükarten, die dürersche Blumenstillleben auf dem Fond zierten und mit wie in Marmor gehauenen Versalienschriften „Gabelfrühstück“, „Appetithappen“ und zum krönenden Abschluss einen „Mokka“ annoncierten. 1958 wurde sogar Holsten-Bier aus Holzfässern an Bord ausgeschänkt. Bis Mitte der 1960er Jahre gab es zudem in der First Class der Boeing 707 eine Stehbar mit Barmixer – Übernahme von und Anlehnung zugleich an das erotisch kodierte Bedeutungspotenzial des „Nachtlebens“ in diversen Etablissements der Großstädte. Selbstverständlich gehörte Rauchen ebenso wie die Bar und die Barbekanntschaft während des Fluges noch zu den nicht hinterfragten Genüssen. Auf Interkontinentalflügen nach Fernost gab es – so suggerieren uns wenigstens zeitgenössische Advertisement-Fotografien der Lufthansa – bereits Mitte der 1960er Jahre asiatische Stewardessen. Natürlich ist von heute aus, mit dem Abstand fast eines halben Jahrhunderts, die Patina solcher Aufnahmen ebenso verführerisch wie

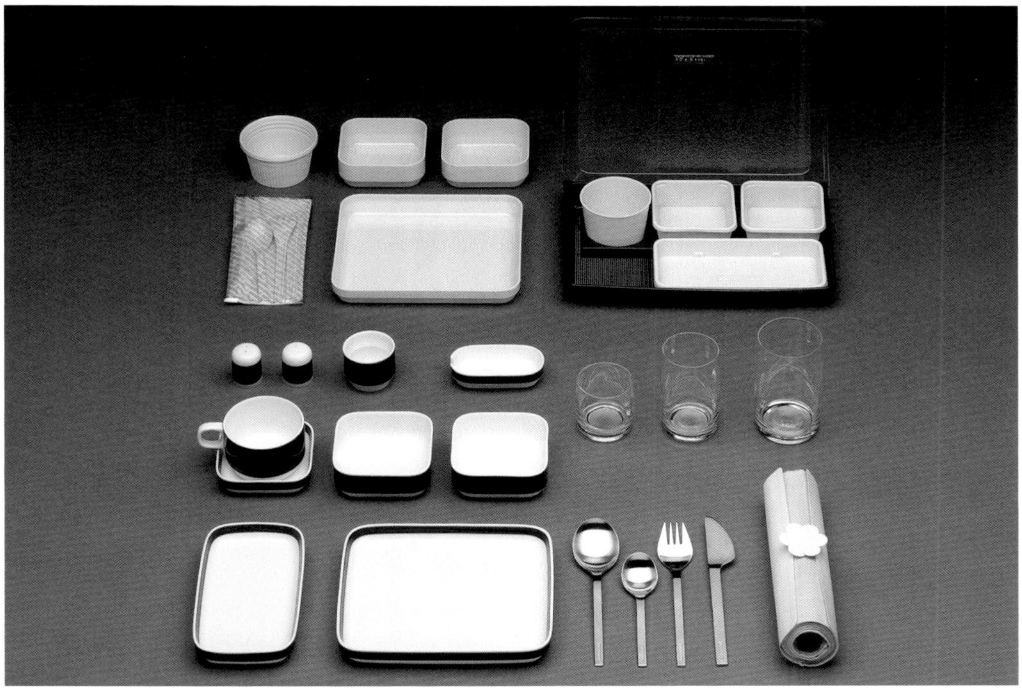

Hans Theo Baumann: Bordgeschirr, First Class, 1971, Hersteller: Hutschenreuther, ab 1976
Hans Theo Baumann: First Class dinnerware, 1971, manufacturer: Hutschenreuther, as of 1976

The Senator breakfast setting was created by Hans Theo Baumann in 1963. It exhibited clear geometric shapes and was manufactured by Rosenthal beginning in 1964. Baumann was a graduate of the College of Design at Ulm, just like Hans "Nick" Roericht who, barely five years later, carried out design work for Lufthansa. Baumann celebrated here for the first time the unadorned and sober product vernacular, doing without any ornamentation whatsoever – the trademark of the Ulm designers. The person to most consistently employ this design language was Dieter Rams. For more than forty years he was chief designer for the Braun electronics company, initially working closely with the college at Ulm. Thus the famous "SK 4" radio receiver with an acrylic cover (nicknamed "Snow White's coffin") was created in cooperation between Dieter Rams and Hans Gugelot at Ulm. And it was hardly a matter of chance that the first in-house designers for Lufthansa all came from the Braun company. But the 1960s were also the decade of brightly colored, optimistic pop art. Hans "Nick" Roericht's plastic dinnerware for Lufthansa was a reflection of this colorful pop culture, so very contrary to the almost ascetic sobriety of the college at Ulm. An ironic bon mot describing the "fiery mouse gray of the Ulm School" still makes the rounds in designer circles. Roericht's design included orange and sunny yellow and was used in the Tourist Class as of 1967/68. The Luran plastic resin employed here was a BASF development. By way of contrast, the more organic "pop art" bowls, which Roericht had developed between 1961 and 1965 in cooperation with Franco Clivio and Michiko Asaka, never got beyond the prototype stage. Hans Theo Baumann then returned to the fore with rectangular plastic serving trays, rounded at the corners and featuring beaded edges to keep dishware and cutlery from sliding off the tray. He had already designed this tray, later manufactured by Hutschenreuther, in 1971. But the 1970s, in spite of the functionalist attitudes which continued to prevail, were also the decade in which lifestyles began to

nostalgisch – wir ahnen, besser, wir wissen: Diese Zeiten der zur Schau gestellten Grandezza des Flugerlebnisses sind im Zeitalter des Massentourismus unwiderruflich vorbei. Heute sitzen agile Jungmanager eifrig über ihre Laptops gebeugt und bestellen eher einen Fruchtsaft als einen Cocktail.

1963 entstand das Senator-Frühstücksgedeck von Hans Theo Baumann in klaren geometrischen Formen, hergestellt von Rosenthal ab 1964. Baumann, ebenso Absolvent der HfG Ulm wie der knapp fünf Jahre später für die Lufthansa entwerfende Hans „Nick" Roericht, zelebrierte hier erstmals die schmucklos-nüchterne, auf alle Ornamentik verzichtende Produktsprache, für die die Ulmer Entwerfer standen. Am konsequentesten übrigens hat diese Produktsprache Dieter Rams als Chefdesigner über 40 Jahre lang für das Unternehmen Braun durchgesetzt, am Anfang mit intensiven Arbeitskontakten zur HfG Ulm. So entstand etwa der berühmte Radioempfänger „SK 4" mit Acrylglasdeckel (der deshalb den Spitznamen „Schneewittchensarg" erhielt) in Zusammenarbeit zwischen Dieter Rams und dem Ulmer Hans Gugelot. Und wohl kaum zufällig kamen die ersten fest angestellten Lufthansa-Designer aus dem Unternehmen Braun. Aber die 1960er Jahre waren nicht zuletzt auch das Jahrzehnt der bunten, optimistischen Pop-Art. Ein Reflex dieser farbintensiven Popkultur, die so konträr zur fast asketischen Nüchternheit der HfG Ulm stand – in Designerkreisen kursiert bis heute das ironische Bonmot vom „feurigen Mausgrau der Ulmer Schule" – war Hans „Nick" Roerichts Kunststoffgeschirr für die Lufthansa, u. a. in Orange und Sonnengelb, das ab 1967/68 in der Touristenklasse eingesetzt wurde. Den hier verwendeten Kunststoff Luran hatte die BASF entwickelt. Jene organoiden „Pop-Art"-Schalen, die Roericht in Zusammenarbeit mit Franco Clivio und Michiko Asaka zwischen 1961 und 1965 entwickelt hatte, blieben dagegen Prototypen. Ab 1976 kam erneut Hans Theo Baumann zum Zug mit rektangulären, an den Ecken abgerundeten Kunststofftabletts mit einem das Tablettgut vor dem Herabrutschen schützenden Außenwulst. Entworfen hatte er diese dann von Hutschenreuther produzierten Tabletts bereits 1971. Aber die 1970er Jahre waren bei allem weiterhin gültigen Funktionalismus auch das Jahrzehnt der auseinanderdriftenden „Lebensstile". Die Gewissheit von einem allein gültigen „Königsweg der Moderne" wurde brüchig. Man mag in jenem aus Keramik gefertigten „Bauernservice" mit gründerzeitlicher Ornamentik, das die Lufthansa in den 1970er Jahren zusätzlich einsetzte, einen Beleg für die damals grassierende Nostalgie sehen. In der First Class der Lufthansa gab es dann entsprechend nostalgische „typisch deutsche" Speisen wie etwa Pfälzer Weinvesper, Heringstöpfchen, Nürnberger Bratwürstchen und Bommerlunder Schnaps in eigenen chamoisfarbenen Gefäßen mit dunkelblauen Dekorschlieren. Da war das Holland des 17. Jahrhunderts formal nicht so weit weg. Indem so das Speisen über den Wolken zum „Dorfgasthaus" romantisiert wurde, wurden damit auch die Großküchenbedingungen, also die industriell rationalisierten Abläufe, ausgeblendet. Da war das Einwegtablett „Eurobox" mit in einzelnen Schalen separiert zellophanierten Menüteilen, das ab 1979 eingesetzt wurde, ehrlicher in seinem Bezug zur Gegenwart und deren Charakteristik.

Die sicherlich größte, auch journalistische Resonanz hatte dann aber 1985 jener auf sieben eingeladene Teilnehmer beschränkte Wettbewerb zur Einführung eines neuen Lufthansa-Bordgeschirrs, den der erste der beiden Finalisten Wolf Karnagel und Dieter Rams für sich entscheiden konnte. „Es geht um die Lufthansa, um die Ausstattung des Bordservices, um das vieldiskutierte Besteck ebenso wie um alles andere, was die Küche an Bord an Utensilien gebraucht. Allein hat er gekämpft, in nur eineinhalb Jahren entstanden mit Hilfe zweier Studenten und eines Freundes rund 600 Prototypen. Und er wusste, dass er in Dieter Rams und seiner Design-Abteilung einen starken

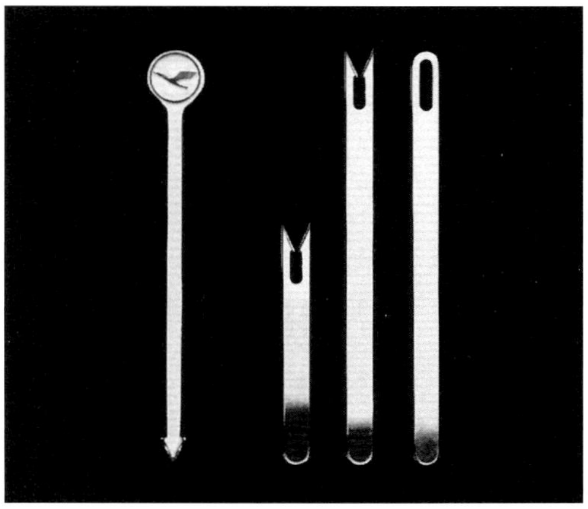

Wolf Karnagel: Cocktail Pieker, 1985
Wolf Karnagel: Swizzle sticks, 1985

Wolf Karnagel: Käsemesser, Universalmesser, Tranchierbesteck, 1985
Wolf Karnagel: Cheese knife, all-purpose knife, carving set, 1985

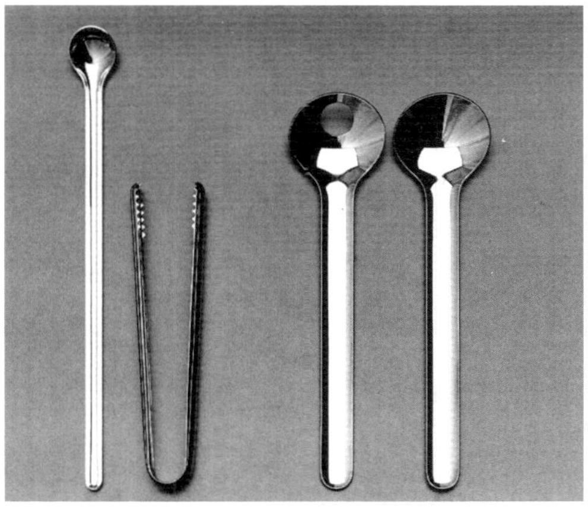

Wolf Karnagel: Cocktaillöffel, Eiszange, Salatbesteck, 1985
Wolf Karnagel: Cocktail spoons, ice tongs, salad service, 1985

Wolf Karnagel: Bordgeschirr, 1985, Hersteller: Hutschenreuther
Wolf Karnagel: Airline dinnerware, 1985, manufacturer: Hutschenreuther

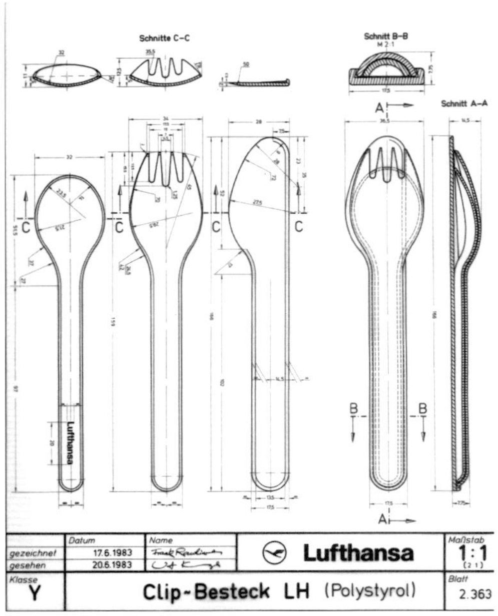


Wolf Karnagel: Entwurf Clip-Besteck LH aus Polystyrol, 1983
Wolf Karnagel: Draft for Lufthansa "Clip" cutlery, made of polystyrene, 1983

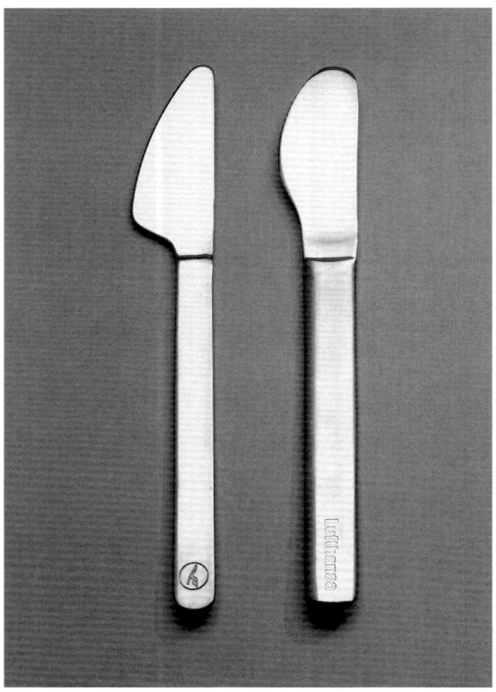


Wolf Karnagel: Messer-Varianten, ab 1985
Wolf Karnagel: Knife variants, as of 1985

Gegner hatte. Das hat ihn nicht verzagen lassen, sondern angespornt. War es der Grund, dass er gewann? ‚Schwer zu sagen. Eines jedenfalls halte ich für einen Vorteil. Ich baue meine Prototypen selbst. Wenn ich sie vorstelle, dann sind sie zuvor wirklich durch meine Hände gegangen. Das ist meine Qualitätskontrolle.'"[4]

Der bereits in den 1960er und 1970er Jahren für Lufthansa-Bordequipment tätige Hans Theo Baumann warf in sowohl gerichtlichen wie journalistischen Auseinandersetzungen Karnagel Plagiatorentum vor. Aber sowohl die Justiz wie auch die Presse hielten Baumanns Vorwürfe für nicht stichhaltig und wiesen sie zurück.[5] Die Debatte drehte sich um Autorschaft und Imitation der neuen Catering-Ausstattung, die im Rahmen des mit dem Winterflugplan 1986/87 generell durchgeführten Relaunches des Gesamterscheinungsbildes der Airline implantiert und von der Schweizer Agentur Zintzmeyer & Lux entwickelt wurde. Für Karnagel stand dementsprechend die sorgfältige Abstimmung mit der neuen Lufthansa-CI im Vordergrund. Dabei wird die Komplexität einer solchen Aufgabenstellung oft unterschätzt. Zu Recht betonte die deutsche Fachzeitschrift *form*:

„Hinter den Kulissen des heutigen Massen-Flugbetriebes wirken Mitarbeiter verschiedener Abteilungen zusammen wie die Räder eines eng verzahnten Getriebes. Aufgabenbereich und Organisation einer jeden Abteilung sind derart festgelegt, dass Veränderungen an einer Stelle sich immer auf andere Bereiche auswirken. Borddienstartikel sind daher in vielerlei Hinsicht von der Betriebsökonomie geprägte Produkte. Geringes Gewicht, knappe, definierte Abmessungen, platzsparende Lagerung, gute Reinigung, Schnelligkeit und Selbstverständlichkeit im Umgang mit ihnen sowie Stoß- und Schlagunempfindlichkeit sind von Betriebsseite die Anforderungen, die an sie gestellt werden. Das Atlassystem, eine Kombination aufeinander abgestimmter Behälter, Wagen, Spül- und Reinigungsmaschinen, das von mehreren Fluggesellschaften gemeinsam genutzt wird, hat die räumlichen Dimensionen etlicher Artikel vorbestimmt."[6]

Die von Karnagel entworfene Table-Top-Ausstattung besteht aus 120 Teilen in den Materialien Glas, Kunststoff, Porzellan und Metall. Fast geometrisch einfache Formen dominieren diese Serie, die auf stapelbaren Rechteck-, Quadrat- und Kreisgefäßen basiert. Die Gläser wirken bei aller formalen Klarheit gedrungen und müssen es wegen der Standfestigkeit wohl auch sein. Am ungewöhnlichsten ist das von Karnagel entworfene, ebenso gedrungene Besteck aus gewalztem Blech: Der kreisförmige Löffel ebenso wie die kurzzinkige Gabel gehen nahtlos in Griffe mit Hohlkehlen über, während die Messerschneide aus ergonomischen Gründen (Schneidewinkel) eine ausladende Kreissegmentform charakterisiert. Allerdings erwiesen sich die Hohlkehlen von Löffel- und Gabelstiel nur bedingt als suppen- und soßentauglich, da die Flüssigkeiten bei der Benutzung manchmal in die Besteckkehlen zurückliefen. Deshalb wurde nach nur einem halben Jahr das Besteck ausgetauscht, zunächst gegen eine flachere Version, ebenfalls von Karnagel entworfen, und dann zugunsten eines vom Fachbereich „Bord Material" der Lufthansa selbst entwickelten Besteckes, das formal sich durchaus an jenes von Karnagel anlehnt, nun aber deutlich noch flachere Griffe aufweist. Trotzdem hatte das ursprüngliche Karnagel-Besteck Avantgarde-Qualitäten. Es vermittelte in seiner durchdachten Funktionalität und trotz oder auch gerade wegen seiner Stapelbarkeit – die einzelnen Teile passen ineinander – einerseits die klassischen Werte der Lufthansa wie technische Kompetenz, Sicherheit und Zuverlässigkeit, andererseits auch die von ihrer neuen Corporate Identity geforderten emotionalen Werte und Assoziationen. Häufig genannte Charakterisierungen wie kindlich, knubbelig, harmlos, präzise, technisch, korrekt, ehrlich bezeichnen die polarisierten Wahrnehmungsmöglichkeiten dieses Entwurfs. Er war ein sehr deutliches Statement eines externen Produktdesigners,

drift apart. The primacy of the "high road of the moderne," valid in every conceivable situation, began to crumble. One might see evidence of the wave of nostalgia which ran rampant at the time in the "country kitchen place setting," made of china and with *fin de siècle* decoration, and used by Lufthansa as an alternate service in the 1970s. At that time nostalgic and "typically German" foods were found in First Class, examples being the Palatinate supper with wine, herring in cream sauce, Nürnberg bratwurst and Bommerlunder schnapps in their own chamois-colored shot glasses with dark blue decoration. Seventeenth-century Holland seemed not so far removed. The romanticizing of dining above the clouds in a "old-fashioned village inn" atmosphere made it possible to ignore the commercial kitchens in the background and their processes, characterized by industrialized rationalization. By contrast, the disposable "Eurobox" serving tray with individual menu items served in cellophane-wrapped dishes – used as of 1979 – was more honest in its relationship to its day and the proclivities of the period.

Certainly the greatest resonance – both in professional circles and in the press – was enjoyed by a 1985 competition held prior to the introduction of new Lufthansa meal service utensils. Participation was limited by invitation to just seven contestants. Of the two finalists, Wolf Karnagel and Dieter Rams, the former carried the field.

"It was a question of Lufthansa, of the meal service appointments, of the much-discussed cutlery and of all the other utensils needed in the galley. He worked on his own and in just one and a half years, assisted by two students and a friend, some 600 prototypes were created. He knew he had a powerful opponent in Dieter Rams and his design department. That did not daunt him but instead spurred him forward. Why did he win? 'That's difficult to say. I did have one advantage, to be certain. I fabricate the prototypes myself. When I unveil them they have actually passed through my hands beforehand. That is my own personal quality check'."[4]

Hans Theo Baumann, who had already contributed to Lufthansa's on-board equipment in the sixties and seventies, accused Karnagel of plagiarism, in both the courts and the press. But neither justice nor journalism felt Baumann's accusations to be justified and thus rejected them.[5] The debate centered upon the authorship or imitation of the new catering equipment, introduced in the course of the general relaunch for the overall image, developed by the Zintzmeyer & Lux agency in Switzerland and adopted by the airline with the winter 1986/87 schedule. In Karnagel's view careful coordination with the new Lufthansa corporate identity was thus in the foreground. The complexity of such an assignment is often underestimated. Quite correctly, the professional journal *Form* in Germany emphasized:

"Behind the scenes in today's high-volume flight operations employees from various departments work together like the gears in a precision-engineered transmission. The tasks and organization in each department are so narrowly defined that changes at any point will always have effects on other areas. Food service items are thus, in many regards, products which bear the mark of operating efficiency. Low weight, compact and closely defined dimensions, space-saving storage, good cleaning properties, speed and ease in handling, and high impact resistance are the demands which operations impose. The 'Atlas' system – a combination of coordinated containers, carts, dishwashers and cleaning machines used jointly by several air carriers – pre-determines the dimensions of many items."[6]

The meal service accessories designed by Karnagel comprised 120 pieces made of glass, plastic, china and metal. Simple, almost geometric shapes dominate this series, which is based upon stackable rectangular, square and circular meal dishes. The glasses, with all their formal clarity, seem rather squat and in fact must be so in the

Wolf Karnagel: Porzellanservice, 1985
Wolf Karnagel: China dinnerware, 1985

Porzellangeschirr, First Class, 1998, Hersteller: Schönwald; Besteck, First Class, 1998; Salz- und Pfefferstreuer, First Class, Hersteller: Palm Design (Auswahl durch Frog Design)
China dinnerware, First Class, 1998, manufacturer: Schönwald; cutlery, First Class, 1998; Salt and Pepper Shaker, First Class, manufacturer: Palm Design (selected by Frog Design)

der selbst im übrigen die aus schwarzem Kunststoff bestehende Besteckfassung favorisierte. Die gesamte Ausstattung besteht aus drei Segmenten: Artikel für die Economy Class, die First Class und Business Class und Servierartikel, mit denen vorwiegend oder ausschließlich die Flugbegleiter umgehen, wie z. B. Cocktail- oder Tranchierbestecke. Karnagel hat immer betont, dass er keine Klassenunterschiede für seine Entwürfe gelten lässt: „Mein bester Entwurf ist das Kunststoff-Clip-Besteck für die Economy-Klasse."[7] Umgekehrt war gerade dieses deutliche Statement in Hinsicht auf die Fluggast-Akzeptanz offenbar nicht unproblematisch, denn vielen Fluggästen erschien die gesamte Serie zu kühl, zu kalt, zu technophil. Die fehlende Suppentauglichkeit allerdings verweist die Fachliteratur in den Bereich der üblen Nachrede:

„Einer immer wieder erzählten Geschichte zufolge mußte eine zweite, etwas flachere Version des Bestecks von Karnagel nur deshalb eingeführt werden, weil sich die Manager an Bord darüber beschwerten, dass ihnen die Suppe in den Ärmel liefe, wenn sie den Arm zum Mund führten. Da es bei der Lufthansa keine Suppe gibt, kann sie aber wohl in den Bereich der Gerüchte verwiesen werden."[8]

Zur Verwendung von Kunststoff bleibt noch zu erwähnen, dass dies nicht nur eine Frage eines demokratischen Selbstverständnisses ist, wie Karnagel so deutlich betont:

„Das Gewicht der Borddienstartikel stellt eine nicht zu vernachlässigende Größe dar. Jedes Gramm muß in die Lüfte gehoben werden und kostet Treibstoff. Die Economy Class stellt bekanntlich die größte Sektion im Flugzeug. Die Gewichte der einzelnen Teile, die dort für den Dienst am Fluggast benötigt werden, multiplizieren sich überproportional. Ein schweres Material wie Porzellan in den dafür erforderlichen Mengen zu transportieren, ist daher abwegig. In der Economy Class wird deshalb Mehrweg-Geschirr aus Kunststoff geflogen."[9] Insgesamt bilanziert die Fachpresse:

„Immer wieder bereichern freudvolle Formen den Ausdruck sachlicher Funktionalität. Seien es Salz- und Pfefferstreuer, Kaviarschaufel oder Fischmesser und -teller, dessen Schwanzflosse der Platz für eine Zitronenscheibe oder besondere Zutaten, wie Meerrettichsahne, sein kann. In der First und Business Class werden nicht nur fertig gedeckte Tabletts serviert, sondern auch Roastbeef vor den Augen der Passagiere tranchiert, Salate frisch angemacht und Käse vom Stück geschnitten ... Bis zum Oktober 1984, als die Lufthansa-Verantwortlichen sich mit 7:1 Stimmen für Karnagels Entwurf entschieden haben, mußten viele Wege eingeschlagen werden, um zu dem dargestellten Ergebnis zu gelangen. Annähernd 600 Modelle sind in dieser Zeit entstanden."[10]

Allein diese Zahl belegt neben allen Feinheiten gestalterischer Überlegungen die logistische Komplexität dieser Bordausstattung: ein Meilenstein im Diskurs miteinander vernetzten Systemdesigns.

Erst wieder mit dem 2001 eingeführten Systemgeschirr „Wing" von Martin Topel und Herb B. Pauser unternahm die Lufthansa erneut vergleichbare Anstrengungen. Demgegenüber wirkt das 1997/98 entworfene First-Class-Besteck von Frog Design wie ein fast nostalgisches Zwischenspiel. Gewissermaßen bourgeois, mit kanellierten Griffen, atmet es geradezu eine bürgerliche Eleganz: Retro-Design mit Anklängen an gründerzeitliches Silberbesteck.

Seit 2001 wird, wie erwähnt, das von Topel und Pauser (Mitarbeit: Ralph Pietruska, Eckhard Ruoff) entworfene Economy-Class-Bordgeschirr „Wing" eingesetzt, dessen Speiseschalen – jeweils in einem Stück aus Kunststoff gefertigte Teile – durchaus spitze Ecken mit wahlweise konkaven oder konvexen Bögen zur Gesamtform verbinden. Damit ist ihre absolut rutschfeste Verkettbarkeit auf dem Tablett gesichert. Die gemäßigt weiche Formensprache zeigt viele durchdachte Details, die den Bedingungen eines

interest of stability. The most unusual of the group is the cutlery designed by Karnagel, which is also stumpy and is made of rolled sheet metal. The circular spoon and the short-tined fork make a seamless transition to the handles which feature a groove on the upper side whereas for ergonomic reasons (cutting angle) the knife blade features the elongated segment of a circle. However, the grooves in the handles for the forks and spoons turned out to be of limited suitability for soups and sauces since liquids sometimes flowed back into this groove during use. That is why after just one year the cutlery was first replaced with a somewhat flatter version, also of Karnagel's design, and ultimately with a cutlery design which was developed by the Lufthansa "cabin materials" department. This final concept most certainly quotes Karnagel in formal terms but now exhibits significantly flatter handles. In spite of all this, the original Karnagel cutlery certainly had avant-garde qualities. Because of its well thought out functionalism and in spite of – or perhaps exactly because of – its stackability (the individual parts nest one inside another) it projects on the one hand the classic Lufthansa values such as technical expertise, safety and reliability but on the other hand the emotional values and associations demanded by the new Lufthansa corporate identity. The design was often characterized as juvenile, chunky and inane but also precise, technical, correct and honest, showing the highly polarized ways in which the concept could be perceived. This was a very clear design statement made by Karnagel, an outside product designer who himself, by the way, favored the cutlery concept using black plastic. The complete service comprises three distinct segments: items for Economy Class, those for Business and First Classes, and serving pieces reserved primarily or exclusively for use by flight attendants – such as barware and carving sets. Karnagel had always stressed that he made no class distinctions in his designs. "My best design is the plastic clip cutlery for the Economy Class."[7] Conversely, it was exactly that clear and unequivocal statement, mentioned above, which apparently created problems in regard to passenger acceptance. In the view of many fliers the entire series appeared to be too cool, too cold, too technical. The professional literature consigns the alleged difficulties with spooning soup to the category of near-slander.

"According to one story told again and again, a second and somewhat flatter version of Karnagel's cutlery had to be introduced simply because managers flying in the premium classes complained that soup ran down their sleeves when they lifted the spoon. Given the fact that Lufthansa has never served soup, this anecdote can be relegated to the realm of urban myth."[8]

As regards the use of plastic, it remains to be mentioned that this choice is not just a matter of democratic attitudes, as Karnagel emphasized so clearly.

"The weight of the items used for food service represents a magnitude which may by no means be neglected. Every gram has to be lifted and that costs fuel. It is a well-known fact that the Economy Class is the largest section in the aircraft. The weight of the individual service for each passenger there is multiplied many times over. Transporting a heavy material such as china in the quantities required for this purpose is thus not justifiable. That is why reusable plastic dinnerware is employed in the Economy Class."[9] The trade press drew up the following general balance sheet:

"Again and again cheerful shapes soften the hard edge of sober functionalism. Be it salt and pepper shakers, a caviar spoon or a fish knife and fish-shaped plate, the tail fin of which provides space for a slice of lemon or special ingredients such as horseradish cream. Meal service in First and Business Classes is not just limited to standardized trays which are completely finished and ready to serve; a roast of beef might be also carved right at the passenger's seat, salad freshly dressed and cheese cut from the

Topel und Pauser: Bordgeschirr „Wing", 2001, Hersteller: Allit; Tablett: Henrichsmeier
Topel und Pauser: "Wing" dinnerware, 2001, manufacturer: Allit; tray: Henrichsmeier

leichten Fluggeschirrs souverän Rechnung tragen. Die Entwicklungszeit betrug zwei Jahre. Die wichtigste Designvorgabe war die Wirtschaftlichkeit in den komplexen Catering-Abläufen. Beim Tablett ging es dann buchstäblich um Millimeter. 2002 hieß es anlässlich der Verleihung eines Designpreises Industrie Forum Hannover für dieses Bordgeschirr:

„Das Economy-Class-Bordgeschirr Wing berücksichtigt in seiner Konzeption die komplexen Logistikabläufe und wirtschaftlichen Aspekte der Luftverkehrsbranche, wie Lagerung, Tabletteindeckung, Spülbandbestückung, Reinigungs- und Ordnungsvorgänge usw. Die Tabletts erhielten vergrößerte und weichere Griffzonen, die Tablettkupplung wurde optimiert, damit sich die Tabletts in den Servier- und Transportwagen nicht mehr übereinander schieben. Die Geschirrteile bestehen aus Kunststoff (SAN), der sich durch eine bewusst keramische Anmutung auszeichnet. Das Tablett wird aus schlagfestem ABS gefertigt. Durch das verbesserte Stapelverhalten gewährt das Bordgeschirr Wing eine Platz- und Kostenersparnis für das Material von 20–30 Prozent. Trotz der komplexen Systemanforderung bei Gebrauch und Fertigung blieb die ästhetische Qualität erhalten." [11]

Mehr als die konventionelleren Tassen wirken die Schalen leicht fernöstlich und reagieren damit auf globale Kulturkreis-Vermischungen, die ihren Niederschlag ja nicht zuletzt in der gesamten deutschen Esskultur und logischerweise auch in dem dann eingesetzten Equipment gefunden haben. So wie die deutsche Gastro-Kultur in den 1960er Jahren die Pizzerias, in den 1970er Jahren die jugoslawischen und griechischen Lokale integrierte, so seit den 1980er Jahren chinesische, japanische und koreanische Restaurants. Insofern ist es nur konsequent, wenn sich diese multikulturelle Durchmischung dann auch im Design, z. B. eines Flugzeug-Bordgeschirrs, spiegelt. Die jüngste Ergänzung des gegenwärtigen Topel-Pauser-Geschirrs vom Frühjahr 2004, nur innereuropäisch genutzt, sind Porzellan-Essschalen mit hitzebeständigen Aluminiumfoliendekkeln, nochmals leichter, mit konvexen weichen Spannungsbögen, hygienisch wie eine Chip-Fabrik, aber durchaus auch mit dem begrenzten Charme einer Ölsardinendose. Auch hausinterne Designer der Lufthansa haben sich vom Japonismus von Topel und Pauser beeinflussen lassen. Seit Sommer 2004 kommt die zweite, achtteilige, nun aus

Jo Laubner: Bordgeschirr „Avanti", First Class, Kanne, 1990er Jahre, Hersteller: Schönwald
Jo Laubner: "Avanti" dinnerware, First Class, coffee pot, 1990s, manufacturer: Schönwald

wheel … By October of 1984, when those responsible at Lufthansa voted by a margin of 7 to 1 to adopt Karnagel's design, many roads had had to be traveled in order to reach the results. Nearly 600 models were created in the process."[10]

This figure alone, in addition to all the details of design considerations, demonstrates the logistical complexity of these on-board accessories. This was a milestone in discourse among closely interlinked system designs.

It was not until 2001 that, in introducing the "Wing" system of dinnerware by Martin Topel and Herb B. Pauser, Lufthansa once again undertook comparable efforts. The First Class dinnerware designed by Frog Design in 1997/98 seems, by comparison, almost like a nostalgic interlude. To a certain extent bourgeois, with fluted handles, it radiates a feeling of traditional elegance: retro design with echoes of *fin de siècle* silverware.

As previously mentioned, the Economy Class "Wing" tableware, designed by Topel und Pauser (in collaboration with Ralph Pietruska and Eckhard Ruoff) has been in use since 2001. Its plastic meal dishes which, in spite of their appearance, are molded as a single piece, join sharp corners with concave or convex surfaces to create the overall shape. This ensures absolutely slip-proof interlocking on the tray. The soft visual language employed here shows many carefully considered details which skillfully take account of the parameters for lightweight airline tableware. Two years were spent on development. The most important design specification was economy in the complex catering procedures. The layout of the tray was calculated down to the very last millimeter. When one of the design prizes sponsored by the "Industrie Forum Hannover" in 2002 was awarded for this aviation tableware, the jury noted:

"In its overall conception the Wing tableware for the Economy Class takes account of the complex logistic procedures and economic aspects of the air travel industry. These include storage, filling the trays, loading industrial dishwasher conveyor belts, cleaning and stowage procedures, and the like. The serving trays were fitted with larger and softer grip areas; the interlocking edges were optimized so that trays would no longer slide one over another inside the service carts. The dishware is made of plastic (SAN) which very deliberately simulates the look of fine china. The serving tray is made of high-impact ABS. Improved stackability of the Wing dinnerware brings about savings of between 20 and 30 per cent in terms of both space and costs. The esthetic qualities were retained despite the complex requirements imposed by the system in terms of both use and manufacturing."[11]

More than the conventional cups, the meal dishes have a slightly Far Eastern flair and thus respond to the global blending of cultures which has been incorporated into German dining culture and then, logically, into the equipment used on the aircraft. The German restaurant trade absorbed pizzerias in the 1960s and in the 1970s Balkan and Greek restaurants; since the 1980s Chinese, Japanese and Korean restaurants have emerged. Thus it is only consistent for this multi-cultural blend to be mirrored in the design of aircraft dinnerware, for example. The most recent addition to the current Topel und Pauser services, introduced in the spring of 2004 and used only on flights within Europe, are china meal dishes with heat-resistant aluminum foil covers, lighter once again, with soft, convex arcs. Their features include not only hygiene worthy of a computer chip factory but also the understated charm of a sardine tin. The designers at Lufthansa have also succumbed to the influence of the Japanophile designs by Topel und Pauser. Since the summer of 2004 the second, eight-piece generation of Topel und Pauser dinnerware, now made of china, is being used on continental and intercontinental routes.

Topel und Pauser: Bordgeschirr, Porzellan, Business Class, 2004
Topel und Pauser: Airline dinnerware, made of china, Business Class, 2004

Porzellan hergestellte Topel-Pauser-Geschirrgeneration auf Kontinental- und Interkontinentalstrecken zum Einsatz.

Geradezu eine verführerische Eleganz dagegen, mit Anklängen an die Streamline-Ära der 1930er Jahre, strahlt eine Porzellanserie der Produktlinie „Avanti“ aus, die Jo Laubner 1992 für Schönwald entworfen hat und die von 1997 bis 2004 in der First Class der Lufthansa eingesetzt wurde. Seit 2004 sind einige Produkte dieser Serie in der Business Class Intercontinental im Einsatz. Und so schließt sich der Kreis der gastronomischen Bordartikel der Lufthansa vom frühen „Modell 21“ von Don Wollance in organischen Formen der 1950er Jahre zu dieser ebenso organisch entworfenen Porzellanserie im „Retro-Futurismus-Stil“ der Gegenwart.

Noch Zukunftsmusik dagegen sind jene neuen Konzepte für das Essen über den Wolken, die Studenten der Hochschule für Gestaltung in Offenbach in Zusammenarbeit mit der LSG Sky Chefs erarbeiteten. Zwei dieser Konzepte wurden im Februar 2004 in Nizza im Rahmen der IFCA (International Flight Catering Association) ausgestellt. Im Erläuterungsbericht zu den Entwürfen heißt es:

„Essen im Flugzeug, in dieser Zwischenwelt verschiedener Orte und Kulturen, hoch über der Erde, und in einer Zeit, die irgendwo zwischen nicht mehr und noch nicht liegt, soll ein Abbild dieser außergewöhnlichen und besonderen Situation sein. Zwar gehören Frische, Qualität und eine ansprechende Präsentation zu den grundlegenden Aspekten, die bei jeder Mahlzeit erfüllt werden müssen, um die Zufriedenheit des Fluggastes zu garantieren, doch sollte Essen im Flugzeug sich durch mehr auszeichnen … Der besondere Moment, das Erlebnis, soll sich in den Mahlzeiten, in der Wahl der Zutaten und in ihrer Zubereitung sowie der Darreichung wiederfinden. Das Essen soll für den Gast zum besonderen Ereignis auf der Reise werden. In den Entwürfen finden sich

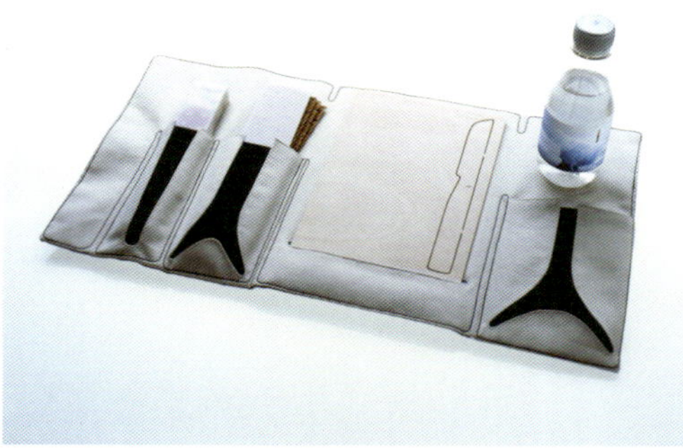

Saftbox
Saftbox

Soupfood
Soupfood

New Trend
New Trend

Basic
Basic

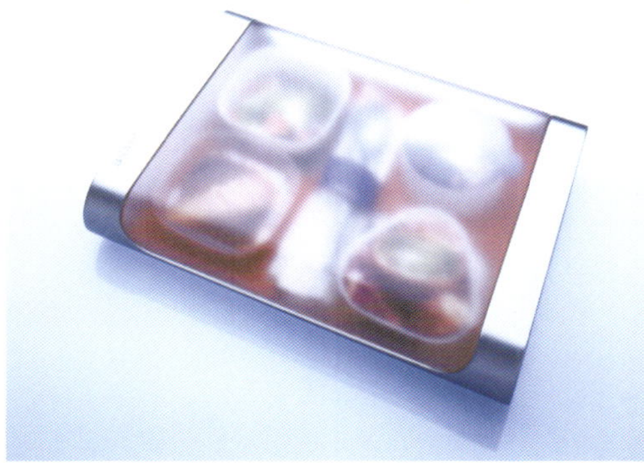

Options
Options

Projektstudie HfG Offenbach, FB Produktdesign mit LSG Sky Chefs: „Essen über den Wolken", 2004/05 (Projektteilnehmer: Christine Bachmann, Andres Hielscher, Nico Hopp, Daniel Knies, Folkert Lübben, Sebastian Lübbert; Betreuer: Peter Eckart)
Project study by the Offenbach College of Design, Product Design Department, in collaboration with LSG Sky Chefs: "Dining above the clouds," 2004/05 (participants: Christine Bachmann, Andres Hielscher, Nico Hopp, Daniel Knies, Folkert Lübben, Sebastian Lübbert; mentor: Peter Eckart)

A dinnerware series from the "Avanti" product line, designed by Jo Laubner for Schönwald in 1992 and used in Lufthansa First Class from 1997 to 2004, radiates a seductive elegance with touches of the "streamline" era of the 1930s. Some of the products from this series have been used in intercontinental Business Class since 2004. And thus the on-board meal service items at Lufthansa have come around full circle – from the early "Model 21" by Don Wollance in the organic shapes of the 1950s to this series of china, just as organic, in today's "retro-futurist" style.

Ideas for the future, by comparison, are those new concepts for dining in the skies which students at the Design College in Offenbach worked out in collaboration with LSG Sky Chefs. Two of these concepts were unveiled in Nice in February of 2004, under the purview of the IFCA (International Flight Catering Association). The explanatory report accompanying the designs says:

"Dining in an airliner, in this world between far-flung places and cultures, high above the earth and in a time which is somewhere between no longer and not yet, is to be a replication of this extraordinary and special situation. Freshness, quality and an appealing presentation are of course among the basic aspects which have to be satisfied in every meal in order to guarantee passenger satisfaction. But meals in an airplane should be distinguished by more than this … The special moment, the experience should be reflected in the meals, in the selection of ingredients, in their preparation and in their presentation. Dining should be a special event for guests during their journey. Various aspects of flight are found in the designs. Lightness, reflection and

verschiedene Aspekte des Fliegens wieder. Leichtigkeit, Besinnung und Erholung werden ebenso zum Zeichen für ein neues Verständnis von Essen über den Wolken wie Anlehnungen an Geschwindigkeit, Technologie und Fortschritt. Neue Materialien und Formen verbinden sich zu Konzepten, in denen sich auf besondere Weise der Reiz des Fliegens widerspiegelt."[12]

Man darf gespannt sein, ob sich die Lufthansa zur Übernahme eines oder mehrerer dieser neuen Konzepte entschließen wird.

Besonderen Wert hat die Lufthansa auch immer auf die grafische Dignität ihrer Speisekarten für die Bordmenüs gelegt. Selbst noch das Schrift- und Bilddesign für die beigegebenen Kleinpackungen, etwa für Salz, Pfeffer, Zucker, Käse-, Tee- oder Salzgebäck, wie auch die Papier- bzw. Stoffservietten entsprachen den jeweils gerade gültigen Gestaltungsrichtlinien. Im Rahmen der ersten umfassenden Corporate Identity der Lufthansa, 1962 von Otl Aicher, waren diese Kleinpackungen rein typografisch gestaltet, klassisch funktional, mit der nüchternen Evidenz eines immer gleich bleibenden Schriftenkataloges. Dies änderte sich Mitte der 1980er Jahre gravierend. Nun gab es Speisekarten mit vollflächigen Farbfotografien auf dem Cover, etwa von Avocados oder Artischocken, wobei diese Früchte jeweils dicht an dicht als Serie einen spannungsreichen visuellen Dialog zwischen aufgeschnittenen und geschlossenen Exemplaren zeigen. Auch die kleinen Kunststofftragetaschen für sogenannte „Lufthansa-Snacks" gehorchten diesem grafischen Prinzip. Parallel dazu gab es in den 1980er Jahren „Patchwork"-Cover mit briefmarkengroßen, seriell angeordneten Fotografien von historischem Steingut, gründerzeitlichem Messinggeschirr, altdeutschem Mobiliar und Fachwerkfassaden, die etwa auf dem Flug von Frankfurt am Main nach New York zum Einsatz kamen und so nicht ohne Nostalgie deutsche Identität über den Atlantik trugen. Ein anderes Konzept der 1980er Jahre favorisierte Speisekarten mit freigestellten Abbildungen von Krebsen, Artischocken, Maiskolben, Paprikaschoten, Avocados, Zwiebeln und Zucchinis, bei denen die Farbvalenz des Hintergrundes die Farbstimmung der Objekte aufnimmt. Die fotografische Exzellenz dieser Aufnahmen lässt unwillkürlich an professionelle „Food-Designer" denken, die mit Hilfe von Glycerin die Lebensmittel buchstäblich auf Hochglanz bringen. Auch wenn die abgebildeten Speisen manchmal nicht den tatsächlichen Menüs entsprechen, sind sie doch ein „Appetitanreger" von hohen Graden. Wie in einem noblen Restaurant am Boden beginnt die „Neugier" auf ein Dinner immer mit seiner visuellen Ankündigung. Andererseits kamen auch wesentlich abstraktere Fotografien zum Einsatz: so etwa in den 1990er Jahren bei den Menükarten der Business Class und der First Class Luftaufnahmen mit im Wasser stehenden Reihen von Reispflanzen oder ein aus großer Höhe fotografiertes Kornfeld. Viele der Kleinpackungen der Mittachtzigerjahre für Zucker, Pfeffer und Salz sowie Erfrischungstücher stammen von Klaus Wille, einem ehemaligen Mitarbeiter Otl Aichers von der Ulmer Hochschule für Gestaltung, der der erste festangestellte „Inhouse"-Designer der Lufthansa wurde. Und so sieht man diesen wunderbar nüchternen, grafisch klaren Entwürfen sehr deutlich die „Aicher"-Schule an.

**1** Keith Lovegrove, *Airline. Identity, Design and Culture*, New York 2000, S. 79.
**2** Doris Schmidt, „Geschirr in der Luft", in: *Zeitgemäße Form. Industrial Design international*, München 1967, S. 91.
**3** Susanne Prinz, *Besteck des 20. Jahrhunderts. Vom Tafelsilber zum Wegwerfartikel*, München 1993, S. 71 f.
**4** Michael Schweer, „Die richtige Kurve ist seine Stärke", *md*, Nr. 2, 1991, S. 49.
**5** Akribisch wird diese Debatte aufgerollt und visuell belegt in dem Artikel von Dieter Willich: „Vom Ursprünglichen und Ähnlichen ... oder die Quadratur des Plagiarius: Baumann gegen Karnagel, Karnagel gegen Baumann, Baumann gegen Willich", *form*, Nr. 130, 1991, S. 64 bis 67.
**6** „‚Freudvolle Formen als Ausdruck sachlicher Funktionalität.' Wolf Karnagel gestaltete für die Deutsche Lufthansa die neue Bordausstattung", *form*, Nr. 116, 1986, S. 25–31.
**7** Zit. nach: Freudvolle Formen ..., a. a. O., S. 27. Weiter heißt es: „Die Idee zu diesem Besteck entstand unter dem Eindruck der immensen Abfallberge, die sich im Flugverkehr anhäufen. Unangetastete Lebensmittel und nur einmal gebrauchte Produkte werden täglich tonnenweise vergeudet. (Im Englischen heißt ‚waste' sowohl ‚Abfall' als auch ‚Vergeudung'.) Das Bemühen, gänzlich auf Einwegartikel zu verzichten, mußte bei näherer Betrachtung der Betriebsabläufe sowohl auf Kurz-, Mittel- als auch Langstreckenflügen aufgegeben werden. Jeder Schritt in diese Richtung läuft auf ein Mehr an Arbeitsgängen hinaus und scheitert spätestens an gewerkschaftlichem Widerstand. Jeder Handgriff scheint tariflich festgelegt. Der Gedanke lag also nahe, ein Einweg-Besteck zu entwerfen, das so schön und über den Flug hinaus nützlich ist, dass es jeder Fluggast gern mitnimmt. Ein Besteck, das neben dem Nutzen für den Fluggast einen wirkungsvollen Werbeträger für die Lufthansa abgeben würde. Dabei kam Karnagel entgegen, dass nur drei Besteckteile erforderlich waren: Messer, Gabel und kleiner Löffel. Eine Kassette entstand, eine Besteckfamilie; das Messer als Boden, die Gabel als Deckel und dazwischen, geschützt, der kleine Löffel."
**8** Susanne Prinz, a. a. O., S. 90.
**9** „Freudvolle Formen ...", a. a. O., S. 28.
**10** „Freudvolle Formen ...", a. a. O., S. 31.
**11** Zit. nach: International Forum Design GmbH (Hrsg.), *Designpreis Industrie Forum Hannover 2002*, Hannover 2002, S. 381.
**12** Daniel Knies, „Flugzeuge im Bauch – Essen über den Wolken", in: Hochschule für Gestaltung Offenbach am Main (Hrsg.), *auch anders – Jahresbericht 2004*, Offenbach am Main 2004, S. 226.

rejuvenation are just as much signs for a new understanding of dining above the clouds as are suggestions of speed, technology and progress. New materials and shapes are joined to create concepts in which the excitement of flying is reflected in a very special way."[12]

One may well be curious to learn whether Lufthansa decides to adopt one or more of these new concepts.

Lufthansa has also always attached special importance to the graphic integrity of its menus for inflight meals. Close attention to detail ensures that the typography and images used for the packets for salt, pepper, sugar, cheese, tea or salted snacks – and for paper or cloth napkins – comply with the design guidelines in effect at the moment. In the course of the first comprehensive corporate identity concept drafted for Lufthansa by Otl Aicher in 1962 these packets used typography only in their design – classically functional and with the sober presence of a family of typefaces which seemed almost immutable. This changed abruptly in the middle of the 1980s. Menus appeared with color photographs on the front cover. Avocados or artichokes might have been shown, for example, in a closely-spaced series of photos documenting the transition from whole fruit to a delicacy sliced and ready for serving – a stimulating and appetizing visual dialog. The small plastic carry bags for the so-called "Lufthansa Snacks" also adhered to this graphic principle. Used in parallel to this concept were "patchwork" menu covers in the 1980s with postage stamp sized photographs, arranged in rows, showing historical crockery, turn-of-the-century brass tableware, antique German furniture and half-timbered facades. These covers were used, for example, on the flight from Frankfurt to New York and thus transported a bit of German identity across the Atlantic, not without a touch of nostalgia. Another concept of the 1980s favored menus with dropped-out illustrations of shellfish, artichokes, corncobs, bell peppers, avocados, onions and zucchinis; the background colors reflected the colors of the subjects themselves. The esthetic quality of the photos unavoidably calls to mind the professional "food designer" who, with the help of a little glycerin, literally brings foods to a high shine. Even though the foods shown did not always correspond with those actually served, they did an excellent job of whetting the observer's appetite. As in an elegant restaurant on the ground, curiosity about a dinner always begins with its visual announcement. On the other hand, far larger numbers of more abstract photographs were used, as well. In the 1990s the menu covers in Business and First Class showed aerial photographs of flooded rows of rice or a wheat field photographed from a great height. Many of the packets used in the mid-eighties – for sugar, pepper, salt and astringent towelettes – were the work of Klaus Wille, a former associate of Otl Aicher at the Ulm Design College and the first "in-house" designer to be hired at Lufthansa. And so one sees the "Aicher school" quite distinctly in his wonderfully sober and graphically clear designs.

**1** Keith Lovegrove, *Airline. Identity, Design and Culture.* New York, London, 2000, p. 79.
**2** Doris Schmidt, "Geschirr in der Luft," in: *Zeitgemäße Form. Industrial Design international*, Munich, 1967, p. 91.
**3** Susanne Prinz, *Besteck des 20. Jahrhunderts. Vom Tafelsilber zum Wegwerfartikel* Munich, 1993, pp. 71 f.
**4** Michael Schweer, "Die richtige Kurve ist seine Stärke", *md*, no. 2, 1991, p. 49.
**5** This debate is documented in great detail and augmented with illustrations in an article by Dieter Willich: "Vom Ursprünglichen und Ähnlichen … oder die Quadratur des Plagiarius: Baumann gegen Karnagel, Karnagel gegen Baumann, Baumann gegen Willich," *form*, no. 130, 1991, pp. 64–67
**6** "'Freudvolle Formen als Ausdruck sachlicher Funktionalität.' Wolf Karnagel gestaltete für die Deutsche Lufthansa die neue Bordausstattung," *form*, no. 116, 1986, pp. 25 to 31.
**7** Quoted in: "Freudvolle Formen …" op. cit., p. 27. He continues there: "The idea for this cutlery grew up under the impression made by the immense mountains of waste which are piled up in air travel. Untouched foods and products used one time only are discarded by the ton every day … The effort to do entirely without disposable items had to be abandoned after closer observation of the operational procedures on short-, medium- and long-distance flights. Every step taken toward reducing waste implied more operations and failed, at the very latest, due to resistance by the unions. Every motion seems to be set down in collective bargaining agreements. And thus it seemed obvious to design disposable cutlery which is so attractive and useful – after the flight has ended, too – that passengers are glad to take it along. This is cutlery which, in addition to utility for the passenger, would also be an effective advertising medium for Lufthansa. Working to Karnagel's advantage here was the fact that only three utensils were required: a knife, a fork and a small spoon. A type of miniature case was created – a cutlery family; the knife as the back, the fork as the lid and in between, protected, the small spoon."
**8** Susanne Prinz, op. cit., p. 90.
**9** "Freudvolle Formen …," op. cit., p. 28.
**10** "Freudvolle Formen …," op. cit., p. 31.
**11** Quoted from: International Forum Design GmbH (ed.), *Designpreis Industrie Forum Hannover 2002*, Hanover, 2002, p. 381.
**12** Daniel Knies, "Flugzeuge im Bauch – Essen über den Wolken," in: Hochschule für Gestaltung Offenbach am Main (ed.), *auch anders – Jahresbericht 2004*, Offenbach am Main, 2004, p. 226.

Speisekarten, Business Class, Mitte 1970er Jahre
Menus, Business Class, mid-1970s

Speisekarten, First Class, 1970er Jahre
Menus, First Class, 1970s

Speisekarten, Speisefolgen á la carte, Mitte 1970er Jahre
Dinner menus, Á la carte selections, late 1970s

Speisekarten, Ende 1970er Jahre
Menus, late 1970s

Speisekarten, Anfang 1980er Jahre
Menus, early 1980s

Speisekarten, Business Class, Ende 1980er Jahre
Menus, Business Class, late 1980s

Speisekarten, First Class, Business Class, Economy Class, Mitte 1990er Jahre
Menus, First Class, Business Class, Economy Class, mid-1990s

OD
Lufthansa
D-ABYA

# Prêt-à-porter der Lüfte

Die Bekleidung der Cockpit-Mitarbeiter und Flugbegleiter ist ebenso wie jene der Stationsmitarbeiter zuallererst Berufskleidung. Sie muss dementsprechend praktisch und widerstandsfähig sein, zugleich aber ansprechend, elegant und modisch. Mode also in Dienst genommen: im Spagat zwischen Couturier und Corporate Culture. Die Ausstrahlung und das Erscheinungsbild des fliegenden Personals sind Faktoren, die auf der Sympathiekurve einer Fluggesellschaft seit je ziemlich weit oben stehen. Entsprechend wird ihnen gestalterische Aufmerksamkeit zuteil. Nun ist ein enger Flugzeuggang nur sehr begrenzt als Laufsteg geeignet, obwohl gerade die Stewardessen-Uniformen bei vielen Fluggesellschaften immer wieder durch Couturiers entworfen wurden und werden. Hardy Amies, Giorgio Armani, Cristóbal Balenciaga, Pierre Balmain, Carven, Christian Dior, Louis Ferraud, Halston, Calvin Klein, Christian Lacroix, Ralph Lauren, Raymond Loewy, Jean Louis, Julian MacDonald, Jean Patou, Emilio Pucci, Mary Quant, Nina Ricci, Yves Saint Laurent, Gabriele Strehle und Valentino sind nur einige der Modeschöpfer, die auch in diesem auf Arbeitsabläufe hin orientierten Feld erfolgreich gewirkt haben. Auch hier wieder vollzieht die Mode wie die anderen Produktfelder beim und ums Fliegen gesellschaftliche Leitbilder, jeweils zeitgenössische Stillagen und wechselnde Erwartungshorizonte nach. Dabei waren und sind Kopfbedeckungen, Halstücher, Blusen, Kostüme, Taschen und Schuhe bei den Damen jeweils ebenso aufeinander abgestimmt wie bei den Herren Anzüge, Krawatten, Westen, Pullover, Hemden und Schuhwerk. Schon ein flüchtiger Blick in jeweils zeitgleiche Bekleidungskataloge großer Versandhäuser zeigt die nahe Verwandtschaft der jeweiligen Ästhetik: Auch die Models ähneln sich bis in die Frisuren, das Make-up und die Körpersprache hinein. Auch hier wieder partizipiert der Gestaltungsdiskurs an zum Fliegen externen Attraktivitätsfeldern: der Gastronomie und Hotellerie gehobener Klasse mit ihren Empfangs-, Hotelpagen- und Service-Outfits, den Chefetagen der großen Banken und Dienstleister wie etwa Versicherungen, Boutiquen und Modehäuser und deren Corporate Clothing, nicht zuletzt aber und seit je am Autoritätsumfeld der Marine und des Militärs. Mit einem allerdings gravierenden Unterschied: Die Bekleidung des Flugpersonals, ob nun am Boden oder in der Luft, hat immer auch und zuallererst die Corporate Identity der jeweiligen Airline zu transportieren – in der Farbgebung, der Anmutung, der emotionalen Positionierung, aber auch in den Rangabzeichen und Schmuckstücken. Dies hat für die Lufthansa im Rahmen ihrer bis dahin so nicht gekannten Corporate-Identity-Profilierung als erster Otl Aicher erkannt und umgesetzt. Man(n) erwartet von Stewardessen nicht in erster Linie technische Kompetenz wie etwa vom Flugkapitän oder seinen Bordingenieuren, sondern eher fachliche und soziale Kompetenz, etwa bei der Kinderbetreuung im Flugzeug. Zudem ist gastronomische Kenntnis wichtig: Selbstverständlich sind auch die Flugbegleiterinnen wie ihre männlichen Kollegen für „Emergency"-Situationen, seien sie technischer oder psychologischer Natur, bestens ausgebildet und vorbereitet.

Der folgende Überblick zeigt in chronologischer Folge die sechs bisherigen Generationen der Uniformen, die modisch jeweils auf der Höhe der Zeit befindlichen Outfits der Lufthansa. Erst ab 1988 kommen Uniformen für Pilotinnen hinzu.

Zwischen 1955 und 1965 kam ein marineblaues Kostüm mit gleichfarbigem Jockey-Käppchen und weißer Bluse zum Einsatz, aber ebenso hochgeschlossene, gestreifte Blusen mit kleinem Bubikragen oder weiße Blusen mit offenem Pellerinenkragen. Natürlich war in dieser Zeit der Pariser Couturier Christian Dior mit seinen diversen,

Teilnehmer des ersten Flugbegleiter-Lehrgangs, 1955
Participants in the first flight attendants' training course, 1955

Seite 144: Werner Machnik: Uniformen Flugbegleiterinnen, 1970–1978
Page 144: Werner Machnik: Stewardess' uniforms, 1970 to 1978

# Ready-to-wear in the air

The uniforms worn by both flight and cabin crews, just as those for employees on the ground, are first and foremost working clothes. Thus they have to be as practical as they are durable but at the same time attractive, elegant and fashionable. Fashion is thus put to use in the dichotomy between *haute couture* and corporate culture. The impression and appearance projected by the entire airborne staff are traditionally high on the scale of factors which determine an airline's empathy level. Appropriately, considerable attention is devoted to fashion design. It is quite true that a narrow aircraft aisle is hardly suitable for use as a fashion show runway – even though flight attendants' uniforms for many airlines have often been and will continue to be crafted by well-known couturiers. Hardy Amies, Giorgio Armani, Cristóbal Balenciaga, Pierre Balmain, Carven, Christian Dior, Louis Ferraud, Halston, Calvin Klein, Christian Lacroix, Ralph Lauren, Raymond Loewy, Jean Louis, Julian MacDonald, Jean Patou, Emilio Pucci, Mary Quant, Nina Ricci, Yves Saint Laurent, Gabriele Strehle and Valentino are just a few of the fashion designers who have been successful in this special field with its orientation toward the needs of the working world. Fashion, similar to other product sectors associated with flying, has followed society's ideals, the styles in vogue at any given time, and changing expectation horizons. Head coverings, scarves, blouses, suits, purses and shoes for the women were and are coordinated one with the other in just the same way as the suits, ties, vests, sweaters, shirts and shoes worn by the men. Even a brief glance at the clothing catalogs published by large mail-order houses in the same time periods shows the close relationship with contemporary esthetics. The models seen there are similar in regard to hairstyles, make-up and body language. Once again there is a dialog in design with sectors which are in fact outside the realm of aviation: top-flight hotels and restaurants with their uniforms for receptionists, bellboys and service personnel, the executive suites not only of the large banks and insurance companies, but also of boutique and retail clothing chains together with their corporate clothing concepts and, last but not least – and from the very outset, the air of authority imparted by the uniforms found in the navy and the other armed forces. There is, however, one significant difference: clothing for airline personnel, whether on the ground or in the air, must always and primarily convey the airline's corporate identity – in terms of color, pleasant appearance, emotional positioning, and also in accessories and the insignia of rank. Otl Aicher was the first to recognize and implement all this for Lufthansa in the course of his defining a new corporate identity, an effort not realized in such breadth up to that time. Passengers do not expect flight attendants to have the same degree of technical expertise as the pilot, copilot or engineer. Instead they expect professional and social competence in, for instance, tending to children's needs in the air. Knowledge of foods and beverages is also important. And it goes without saying that each and every flight attendant is superbly trained and prepared to deal with emergency situations, be they of a technical or psychological nature.

The following review shows, in chronological sequence, the six past generations of uniforms, each of which carefully followed the fashions of the particular time. It was only in 1988 that uniforms for female pilots were added.

Between 1955 and 1965 a navy blue suit and a jockey cap in the same color were used in combination with either a white blouse or a striped blouse with a small round collar or white blouses with open pelerine collars. Naturally at this time Parisian de-

Uli Richter: Stewardessen-Uniformen, 1955–1965
Uli Richter: Stewardess' uniforms, 1955–1965

Uli Richter: Stewardessen-Uniformen, 1955–1965
Uli Richter: Stewardess' uniforms, 1955–1965

jeweils auf verschiedene „Silhouetten" bezogenen Kleiderlinien formprägend. Als Variante gab es ab Mai 1961 diese Kombination auch in Petrolblau als sogenanntes „Tropenkostüm". Die enge Linie mit anliegendem Oberteil, langem Revers und runden Schultern, das in der Taille geknöpft wurde und ab 1955 bei der Lufthansa zum Einsatz kam, wurde bereits 1948 von Christian Dior erfunden, ebenso 1956 die sogenannte „Pfeillinie" mit kurzem Bolero-Jäckchen, das die hohe Taille eines Festkleides überspielte. Aber auch in der Hüfte gegurtete Kleider mit angeschnittenen Ärmeln in jener engen, auch „Bleistiftlinie" genannten Kontur à la Dior mit begrenzt schwingendem Rock wurden von der Lufthansa verwendet. Die Flugbegleiter trugen Anzüge mit breitem Revers und dünnen Krawatten in der Mode dieser Jahre. Selbst bei hochsommerlichen Temperaturen sollten die Stewardessen die Jockey-Kappen und weiße Handschuhe tragen, auch wenn die Bluse schon einmal kurzärmelig sein durfte.

Entworfen hatte diese ersten Nachkriegs-LH-Flugkollektionen der Berliner Modemacher Uli Richter, der ab 1959 ein eigenes Haute-Couture-Haus in Berlin führte. Im Reclams Mode- und Kostümlexikon heißt es: „Richter entwarf die Hostessen-Uniformen für die Deutsche Lufthansa."[1] Bilder von 1955 zeigen Teilnehmer des ersten Flugbegleiter-Lehrgangs, sowohl Stewardessen wie Flugkapitäne. Die Kapitänsmützen sind locker ballonartig gebauscht: kein Hinweis auf militärische Strenge.

Zwischen 1961 und 1963 gab es für die in der Senator-Klasse arbeitenden Stewardessen den sogenannten „Senator-Dress", der aus einem beigen wadenlangen Etuikleid mit vier Pattentaschen, einem abgewandelten Zylinder, einem Mantel sowie weißen Handschuhen bestand.

Ab 1965 wurden marineblaue Kostüme mit sogenannten „Toque"-Hüten, die mit türkisfarbenen Blusen kombiniert wurden, verwendet. Für die damals etwa 700 Stewardessen hatte sie die ehemalige Chefstewardess Ursula Tautz entworfen, die seit 1961 auch die modische Beratung der Lufthansa übernommen hatte. Auch hier gab es wieder ein „Tropenvariation", die komplett in Türkis gehalten war. Die kurzen Jäckchen dieser schon als Mini geschneiderten Etuikleider waren kragenlos und wurden mit einem Bord-â-Bord-Verschluss auf Brusthöhe geschlossen. Der Kragen der offenen Bluse liegt über der Jacke und vermittelt so ein Gefühl relativ lässiger Freizeitbekleidung, ohne formale Korrektheit und Eleganz zu vernachlässigen. In den Jahren zwischen 1965 und 1970 kamen Uniformen zum Einsatz, die ebenfalls Ursula Tautz entwarf: städtische Kostüme mit passenden Boxtaschen und flachen Toque- bzw. Baretthüten, die Kurzhaarfrisuren nahe legten.

„Silberne ‚Dinnerjackets' wurden im Jahr 1967 für die zu dieser Zeit etwa 140 Purser und Erste-Klasse-Lufthansa-Stewards entworfen. Damit sollte der individuellen Fluggastbetreuung eine neue Note gegeben werden. Auf allen Interkontinentalstrecken der Lufthansa tauschten die Herren der Kabinenbesatzung ihre dunkelblaue Uniformjacke von Servicebeginn an bis zur Landeansage gegen das schicke ‚Dinnerjacket', das in Zusammenarbeit mit dem Kelsterbacher Mode-Designer Erwin Klebbe kreiert worden war. In ihm setzte sich eine Tradition fort, die in den Gründerjahren der Lufthansa mit der eierschalenfarbenen Smokingjacke für den exklusiven ‚Senatordienst' (Kabinendienst in der Ersten Klasse) begann."[2]

Zwischen 1970 und 1978 wurde das Marineblau der Uniformen von dem jungen Berliner Couturier Werner Machnik mit blaugelben Halstüchern mit einem lockeren Touch angereichert. Neben dem Stewardessenkleid gehörten zur Uniform eine mit vier schrägen Pattentaschen dynamisierte Jacke, ein Hut, ein Kurzmantel sowie eine Schürze. In diesem ansonsten an den britischen Mini-Stil à la Mary Quant angelehnten Ensemble wirkt die Hutversion sehr deutsch und erinnert an ähnliche Kollektionen von

signer Christian Dior with his lines of clothing, representing various "silhouettes," had considerable impact. One theme introduced in May of 1961 showed this combination in petrol blue, as well as a so-called "tropical costume." This narrow line with its close-fitting top, long lapels and rounded shoulder-line, which was buttoned at the waist and came into use at Lufthansa in 1955, had actually been created by Christian Dior in 1948. In 1956 the so-called "A-line" with the short bolero jacket which mimicked the high waist-line of an evening gown was introduced. Also used by Lufthansa were dresses belted at the hips, with batwing sleeves in a close-fitting contour à la Dior, also known as the "pencil line," with a skirt showing a limited amount of sway. The male flight attendants sported wide-lapel suit jackets and narrow ties, reflecting the fashion of those years. Even at summer temperatures the stewardesses were required to wear the jockey caps and white gloves, though a short-sleeved blouse might be permitted.

Berlin fashion designer Uli Richter who, beginning in 1959, headed up his own *haute couture* studio in Berlin, made up this first post-war Lufthansa flight collection. In the *Reclams Mode- und Kostümlexikon* it says: "Richter designed the hostess uniforms for the Deutsche Lufthansa."[1] Photos dating from 1955 show participants in the first training session for air crews, including both stewardesses and pilots. The captains' caps are almost casual and balloon-like; there is not a hint of martial severity.

Between 1961 and 1963 one found the so-called "Senator uniform" for the stewardesses working in that class. It comprised a beige, calf-length sheath dress with four flap pockets, a hat in the form of a modified cylinder, a topcoat and white gloves.

Beginning in 1965 navy blue outfits including toque hats in combination with turquoise blouses were introduced. They were designed for the 700 stewardesses in the air at that time by Ursula Tautz, the former chief stewardess who, since 1961, had also been responsible for fashion consulting to Lufthansa. Here, again, there was a "tropical" version which was entirely in turquoise. The short jackets for these sheath dresses, tailored as a miniskirt, were collarless and featured an edge-to-edge closure at breast height. The collars of the blouses were worn unbuttoned, outside the jacket, and thus projected a feeling of relatively casual recreational clothing, without neglecting formal correctness and elegance. The uniforms worn in the years between 1965 and 1970 were also designed by Ursula Tautz: urbane costumes with matching box pockets and flat toque caps and berets which were better suited to shorter hairstyles.

"Silver dinner jackets were created in 1967 for Lufthansa's 140 pursers and First Class stewards. Thus a new note was to be given to individual customer care. On all of Lufthansa's intercontinental routes the men in the cabin crew laid aside the dark blue uniform jacket at the beginning of service and donned the stylish dinner jacket. It was worn until the pre-landing announcement. This item was created in cooperation with fashion designer Erwin Klebbe. It revived a tradition which had begun in Lufthansa's early years – the eggshell dinner jacket for the exclusive 'Senator Service' (First Class cabin service)."[2]

Between 1970 and 1978 the up-and-coming Berlin fashion designer Werner Machnik enriched the navy blue of the uniforms with yellow and blue neckerchiefs, adding a more relaxed touch. The uniform included, in addition to the dress for stewardesses, a jacket rendered more dynamic with four angled flap pockets, a hat, a short coat and an apron. This ensemble did, indeed, invoke comparisons with Mary Quant's British "mini" style. The hat, however, was fairly Germanic, reminding viewers of similar collections by Uli Richter of Berlin, who had indeed created the very first Lufthansa fashion line. This combination followed the trends of the times – the optimistic "Swinging Sixties." It also calls to mind André Courrèges' fashions which, in the 1960s, drew attention with their

Ursula Tautz: Sommeruniform, 1965–1970
Ursula Tautz: Summer uniform, 1965–1970

Werner Machnik: Stewardessen-Uniformen, 1970–1979
Werner Machnik: Stewardess' uniforms, 1970–1979

Uli Richter aus Berlin, der ja bereits die erste Lufthansa-Modelinie kreierte. Diese Kombinatorik folgte dem Zeittrend der optimistischen „swinging sixties", lässt aber auch an André Courrèges' Mode denken, die in den 1960er Jahren durch konstruierte Schnitte und geometrischen Aufbau auffiel. Dem erneut dazugehörigen „Tropenkostüm" in Sonnengelb wurde statt des Kurzmantels ein leichteres, ebenso grellgelbes Cape mit blauem Innenfutter beigegeben, das an Superman und Batman erinnert, in jedem Fall aber sich auf Comic-Elemente bezieht. Zusammen mit dem diagonal blau-weiß gestreiften Halstuch gemahnt dieses farbintensive Ensemble an die Pop-Art und ihre Nobilitierung der Banal-Kultur.

„1974 erhält auch das Bodenpersonal der Lufthansa ein einheitliches Outfit. Zusätzlich zum Rock war zum ersten Mal eine flotte, lange Hose, die mit Bluse oder Pullover kombiniert werden konnte, im Sortiment enthalten. Damit waren die rund 2.200 Damen in den Stadtbüros der Lufthansa und an den Flughäfen ihren Kolleginnen in der Luft erstmals eine modische Nasenlänge voraus."[3]

Werner Machnik: Stewardessen-Uniformen, 1970–1979
Werner Machnik: Stewardess' uniforms, 1970–1979

Ab 1979 kamen bis 1987 relativ konventionelle Blazer-Kostüme im Clubstil und auch Hosenanzüge zum Einsatz, dunkelblau natürlich, mit burschikos geknüpften, blau-weißen Halstüchern, entworfen von dem deutschen Bekleidungsunternehmen Jobis. Diese neue Dienstkleidung umfasste in der Grundausstattung insgesamt 33 Teile. Die Blusen dieser Uniformen waren mit dem Kranich-Dekor bedruckt. Es gab sie, jeweils mit Kantstreifen, in weiß und gelb. Die Hosenanzüge waren eine ebenso emanzipatorische wie praktische Geste mit allerdings durchaus größerer Verspätung im Verhältnis zur allgemeinen gesellschaftlichen Entwicklung, hatten sie doch bereits in den späten 1960er, frühen 1970er Jahren Furore gemacht. Die Lufthansa bilanzierte damals:

„Der Stoffverbrauch für die ‚Grundausstattung' der Stewardessen war enorm: 33.000 Meter dunkelblaues Tuch wurden zu Jacken, Röcken, Hosen und Mänteln verarbeitet, die Servicekleider ‚verschluckten' weitere 9.000 Meter Stoff und die Blusen noch einmal

structured cut and geometric character. The tropical uniform once again made its appearance, this time in sunshine yellow, and was complemented not with the short jacket but with a lightweight cape, also in bright yellow with a blue lining – reminiscent of Superman and Batman and in any case referencing comic figures. Together with the blue and white diagonal stripes on the neck scarf, this brightly colored ensemble reflects the influence of pop art and the elevation of pop culture to the mainstream.

"In 1974 Lufthansa ground personnel also received a uniform outfit. In addition to the skirt, there were dashing trousers which could be combined with a blouse or slipover sweater. Thus the 2,200 women in the Lufthansa downtown offices and at the airports were for the first time ever ahead of their airborne colleagues by a nose, at least in terms of fashions."[3]

Seen between 1979 and 1987 were relatively conventional blazer outfits of traditional cut along with pantsuits – in dark blue, of course – with boyishly knotted, blue and white neckerchiefs, all designed by the German clothier Jobis. In the basic version this new uniform comprised a total of 33 pieces. The blouses for these uniforms were printed with the crane emblem. They were available in either white or yellow, with stripes at the edges. The pantsuits represented a gesture which was as much emancipatory as it was practical, even though they appeared late when compared with general fashion trends; pantsuits such as these had caused something of an uproar when they appeared in the late 1960s and early 1970s. Lufthansa summarized at that time:

"The amount of material used for the 'basic uniform' for stewardesses was enormous: 33,000 meters of dark blue fabric was tailored to make up jackets, skirts, trousers and coats. Clothing for meal service 'consumed' a further 9,000 meters of fabric and the blouses took another 45,000 meters. If one adds the more than 3,000 meters for scarves, then the total length would be a good 90 kilometers."[4]

Jobis: Uniformen Bodenpersonal, 1979–1987
Jobis: Ground-personnel uniforms, 1979–1987

Starting in 1987 things became less clear-cut. Appearances fluctuated between casual and classic; the suit in the familiar navy blue with broad lapels was now worn with a white blouse set off with gray pinstripes, combined with numerous accessories such as neckties, bow ties and pocket handkerchiefs in yellow, blue and gray. The fashion designer responsible in this time frame, Jürgen Weiss, also included short-sleeved shirtwaist dresses with a sporty but elegant appearance, with lighter hems at the sleeves and matching belts. He also re-introduced the calf-length skirt. During meal service the flight attendants wore white cuffs, a white collar and a white neckerchief. Again there was a pantsuit with pleated trousers. It was often worn with a long-sleeved white blouse and combined with a ruffled, oversized neck scarf in bright Lufthansa yellow. Here again, as in the previous Jobis collection, a knitted cardigan was part of the ensemble. It mirrored the casual feeling of the "aerobics generation" and the incorporation of the parameters more typical of the leisure society into the professional world. Certainly contributing to the emergence of this more casual attitude was increased employee participation in the selection of the designs.

Jobis: Uniformen fliegendes Personal, 1979–1987
Jobis: Cabin-crew uniforms, 1979–1987

Female pilots were employed by the Lufthansa for the first time in 1988. But it was the "women in the cockpit" who remained outside the fashion fold for a long time. The few women pilots at that time had to make do with altered men's uniforms. It was not until 1991 that uniforms – new and more feminine – were created especially for women at the helm.[5] Found here is the usual soaring crane emblem, similar to the brooches worn by the stewardesses and simply indicating the association with the airline. They were joined by the insignia of rank which up to then had been reserved to men. Beginning in 1987 the range of variations and accessories for the female flight attendants' uniforms was significantly extended. A listing of the model lines between 1987 and 2001

45.000 Meter. Rechnet man die über 3.000 Meter für Schals noch hinzu, ergibt sich eine Stoffbahn von gut 90 Kilometern."[4]

Ab 1987 wurde es dann noch unentschiedener; das Erscheinungsbild schwankte zwischen „casual" und „klassisch": Das Kostüm im bekannten Marineblau mit breitem Reverskragen wurde nun mit weißer Bluse mit grauem Nadelstreifen getragen, kombiniert mit zahlreichen Accessoires wie Neckslips, Fliegen und Einstecktüchern in Gelb, Blau und Grau. Auch kurzärmelige sportlich-elegante Hemdblusenkleider mit hellem Ärmelsaum und korrespondierendem Gürtel und nun wieder wadenlangem Rock ließ sich der für diesen Zeitraum verantwortliche Modedesigner, Jürgen Weiss, einfallen. Zum Servicekleid trug die Flugbegleiterin weiße Manschetten, einen weißen Kragen und ein weißes Tuch. Erneut gab es auch einen Hosenanzug mit Bundfaltenhose, zu dem oft eine langärmelige, weiße Hemdbluse getragen wurde, kombiniert mit einem geruschten, überdimensionalen Halstuch im kräftigen Lufthansa-Gelb. Erneut gehörte wie in der vorherigen Kollektion von Jobis eine Strickjacke dazu; Hinweis auf die Lockerheit einer „aerobic generation" ebenso wie auf die Durchdringung von Parametern einer Freizeitgesellschaft in die Berufsgesellschaft hinein. Sicherlich war diese Lockerheit aber auch einer wachsenden Mitbestimmung geschuldet.

Jürgen Weiss: Erste Pilotinnen in Uniform, 1988
Jürgen Weiss: First women pilots in uniforms, 1988

Ab 1988 wurden erstmals weibliche Piloten bei der Lufthansa eingesetzt. Aber ausgerechnet „die Damen im Cockpit blieben modisch lange außen vor. Die zu jener Zeit wenigen Pilotinnen mussten sich mit umgeschneiderten Herrenuniformen begnügen. Erst 1991 wurde die Arbeitskleidung für die weibliche Welt im Cockpit neu und femininer gestaltet."[5] Hier nun treten zu den üblichen Kranich-Emblemen, wie den Anstecknadeln der Flugbegleiterinnen, die lediglich die Company-Zugehörigkeit avisieren, die bis dahin den Männern vorbehaltenen Rangabzeichen hinzu. Ab 1987 werden die Varianten und Accessoires der Stewardessenausstattung deutlich umfangreicher. Eine Aufstellung der Modelllinien zwischen 1987 und 2001 weist 18 Bekleidungsteile für die Damen und immerhin 13 für die Herren auf. Dazu gehören für die ersten Röcke mit Schlitz oder einer bzw. sechs Falten, gestreifte Kleider, Hosen mit oder ohne Bundfalte, Gürtel, Weste und Strickjacke, Pullunder und Hosenrock, Schürze und ein Trenchcoat; für die Herren ein- und zweireihige Sakkos, Bundfaltenhosen, graue Servicejackets, Cockpit-Strickjacken, blaue Krawatten, halb- und ganzärmelige Hemden, Pullover, Pullunder und Westen und erneut ebenfalls ein Trenchcoat. Allerdings wurden bei dieser Aufstellung auch Uniformteile für den Boden mitgezählt. Die Blusen dieser Kollektion hatte die Designerin Nicole Dürr entwickelt: Sie zierten unterhalb des Kragens am Rücken das Kranich-Emblem, so dass die Stewardess, auch wenn sie einem Gast den Rücken zuwendet, immer noch deutlich auf die Fluggesellschaft verweist. Das intern sogenannte „Gelbsignal" – von der Agentur Zintzmeyer & Lux in jener Zeit dem Unternehmen generell verordnet – zeigte sich auch in dem bereits erwähnten, „Neckslip" genannten krawattenähnlichen Halstuch, das die Stewardessen damals trugen.

Ab 2002 wurde die bis heute aktuelle Uniformlinie für die Damen von Gabriele Strehle mit ihrem Label „Strenesse" entwickelt. 2001, gefragt nach den neuen Uniformen und deren „spirit", sagte sie: „Sie werden so aussehen, dass es zum Berufsbild passt: freundlich und dynamisch."[6] Ihre Modelinie für Lufthansa zeichnet ein bisschen jener „Retro-Futurismus" aus, den Hartmut Esslinger bei seinem Interieur-Relaunch der Lufthansa seit 1998 als Parole ausgab. So erinnert die Kopfbedeckung der Flugbegleiterinnen einerseits an die „Pillbox" bzw. die Toque-Hüte der 1960er Jahre, aber eben nicht nur, sondern auch etwa an orientalische Feze. Der Blazerkostümschnitt ist figurbetont, fast streamline-mäßig mit schräg aufgesetzten Pattentaschen und ebenso diagonalschmissigen Manschettenmarkierungen, die dadurch eben gerade nicht als Rangab-

shows eighteen items of clothing for women and, indeed, thirteen for men. The women could choose between the first skirt with a slit or a skirt with one or with six pleats, striped dresses, trousers with or without pleats at the waistband, belts, vests and knitted cardigans, slipover sweaters, culottes, an apron and a trench coat. The selection for men included single- and double-breasted sports jackets, pleated trousers, gray food-service jackets, cardigans for the cockpit, blue ties, short- and long-sleeved shirts, sweaters, slipover sweaters and vests and, once again, a trench coat. It should be noted, however, that this listing also includes the uniform items intended for ground personnel. The blouses in this collection were developed by Nicole Dürr. They were decorated with the crane emblem just below the collar on the back, so that the stewardess, even when facing away from the passengers, still clearly showed the airline brand. Called the "yellow beacon" inside the company, it was prescribed for almost universal use by the Zintzmeyer & Lux agency. It was also seen on the kerchief previously mentioned, a scarf similar to a necktie which stewardesses wore then.

Beginning in 2002 the women's uniform line which is still current was developed by Gabriele Strehle with her "Strenesse" label. In 2001, when asked about the new uniforms and their "spirit," she said: "Their appearance will match the image of the profession – friendly and dynamic."[6] Her fashion line for Lufthansa demonstrates something of the "retro-futurist" style which Hartmut Esslinger had used as his guideline since 1998 when he revamped the Lufthansa interiors. Thus head coverings for the female flight attendants remind us of the pillbox or toque hats of the 1960s – and not only of them, but to some extent the oriental fez. The cut of the blazer emphasizes the figure, is almost streamlined, with diagonal flap pockets and cuff details which are also placed on the diagonal so that they do not quite look like military insignia. Here the narrow silhouette of the lengthened jacket, together with the toque cap, puts the ensemble in an Indian fashion context but reflects an elegant and urbane European style just the same. It is hardly a matter of chance that the skirt length selected here is known among fashion professionals as "Parisian length." The blouse is in a dignified white or Lufthansa blue, the neckerchief in navy blue, white and yellow, or yellow alone. Consequently the clear use of the classical Lufthansa corporate colors moves once again into the foreground. The men's collection introduced at the same time was created by Etzkorn couturiers. The cockpit uniforms, whether for men or women, were cast in the classical double-breasted cut. Thus the Strenesse line includes the following basic items for a female flight attendant: a coat, two suit jackets, a set of blouses, trousers and skirts (four items), a fabric vest, a knitted item, belt, woolen scarf, handbag and various accessories such as lightweight scarves. The *Frankfurter Rundschau* cast some light upon the logistical requirements involved in a general conversion of the fashion line for a major corporation such as Lufthansa:

"The largest part of the work involved choosing the correct fabric. Gabriele Strehle had in fact selected a very fine material which, however, was too light in weight for everyday work on board. Many members of the cabin crews have their clothing laundered while they are en route. The material would not have long withstood the aggressive dry cleaning techniques used in the USA, for instance. The creative team got together with a fabric manufacturer and developed – using fibers ranging from the wool of New Zealand sheep through to modern Elastan – an entirely new blend comprising 60 per cent wool, 35 per cent polyester, three per cent Elastan and two per cent polyamide. This is a superior fabric which Lufthansa could quite justifiably have had patented. Wearing tests followed, along with additional changes. Strenesse sold the airline the patterns and Lufthansa had to find a general contractor to make up the clothing.

Jürgen Weiss: Uniform Flugbegleiterin, 1987–2001
Jürgen Weiss: Flight attendants' uniform, 1987–2001

Flugspangen und Namensschilder Herren, bis 2001
Pins and name pates men, through 2001

Flugspangen und Namensschilder Damen, bis 2001
Pins and name pates women, through 2001

zeichen wahrgenommen werden. Dabei rückt die schmale Silhouette der gelängten Jacke zusammen mit der Toque-Kappe das Ensemble stilistisch in indische Modezusammenhänge ebenso, wie hier gleichzeitig ein eleganter mitteleuropäischer Citystil gefunden wurde. Nicht zufällig heißt die hier gewählte Rocklänge in der Fachsprache „Parislänge“. Die Bluse vornehm weiß oder im Lufthansa-Blau, das Halstuch marineblau, weiß und gelb bzw. nur in Gelb: mithin erneut die deutliche Benutzung der klassischen Corporate-Farben der Lufthansa. Die zeitgleiche Herrenkollektion stammt aus dem Hause Etzkorn.

Die Cockpit-Uniformen, ob für Mann oder Frau, wurden im klassischen Zweireiher-Look gestaltet. So gehören bei der Strenesse-Linie zur Grundausstattung einer Flugbegleiterin: ein Mantel, zwei Kostümjacken, ein Blusensatz, vier Unterteile (Hosen und Röcke), eine Stoffweste, ein Strickteil, Gürtel, Wollschal, Handtasche und verschiedene Accessoires wie etwa Tücher. In der *Frankfurter Rundschau* wurden die logistischen Anforderungen einer generellen Umstellung der Modelinien eines Großunternehmens wie der Lufthansa beleuchtet:

„Die meiste Arbeit gab es bei der Auswahl des richtigen Stoffes. Gabriele Strehle hatte einen zwar sehr feinen, aber für den Arbeitsalltag an Bord dann doch zu dünnen Stoff benutzt. Viele Bord-Mitarbeiter lassen ihre Kleidung unterwegs reinigen – das Material hätte beispielsweise die aggressiven Reinigungen in den USA nicht sehr lange unbeschadet überstanden. Das Kreativteam setzte sich mit einem Hersteller zusammen und entwickelte von der Wolle neuseeländischer Schafe bis hin zum Elastan-Anteil einen ganz neuen Stoff: Der hat 60 Prozent Wolle, 35 Prozent Polyester, drei Prozent Elastan und zwei Prozent Polyamid. Ein Super-Stoff, auf den Lufthansa ein Patent anmelden könnte. Es folgten Tragetests und weitere Veränderungen. Strenesse hat der Fluglinie den Schnitt verkauft, Lufthansa musste noch einen Generalunternehmer für die Konfektion suchen. Genäht wird die neue Dienstbekleidung in Osteuropa. Die große logistische Frage lautete: Wie bekommen 25.000 Menschen neue Uniformen zugestellt, die ihnen auch passen? Uta Kötting [bei der Lufthansa in der „Projektleitung neue Dienstbekleidung” tätig, erläutert]: ‚Wenn wir unseren Mitarbeitern die Sachen einfach nur zugeschickt hätten, wären hinterher noch viele individuelle Änderungswünsche gekommen.‘ Deshalb wurde eine ehemalige Gruppen-Lounge zu einer riesigen Umkleidekabine umfunktioniert, in der Tausende Teile der neuen Uniformen inklusive aller Zwischengrößen auslagen. Jacke wie Hose: Der Großteil der Mitarbeiter wurde dort regelrecht durchgeschleust und bekam von Anprobe-Station zu Anprobe-Station seine Daten.“[7]

„Am Ende waren rund 600.000 Meter des speziell für Lufthansa entwickelten blauen Stoffes zu 490.000 Uniformen (Röcken, Hosen, Blazer, Westen etc.) verarbeitet.“[8]

Und im gleichen Jahr hieß es in einem allgemeinen Überblick über die gegenwärtige Mode der Flugbegleiter in der *Frankfurter Allgemeinen Sonntagszeitung* zur Lufthansa stilkritisch:

„Während die türkisen und kanariengelben Siebziger-Outfits im Beinbereich nicht viel zu atmen übrig ließen und die Achtziger mit tiefen Kellerfalten in Dunkelblau überreagierten, hat Designerin Gabriele Strehle für die im Mai gestartete Crew-Uniform Dezenz mit Sex-Appeal gemischt: Modellierte Schultern, figurnahe Linien und knielange Röcke demonstrieren Schnittigkeit und Souveränität. Ein kokettes Retro-Element sind die Pillbox-Hüte. Der Popkultur huldigt die Kranich-Linie, indem sie ihre Logo-Präsenz durch drei markante, goldgelbe Streifen bekräftigt, die – jetzt auch für das weibliche Personal – den Rang bezeichnen.“[9]

Nun unterhält, fast wie nationale Streitkräfte, auch die Lufthansa auf ihrer Frankfurter Basis ein Servicecenter für Dienstbekleidung, die sogenannte „Pillbox“. Hier werden die

The new service garments were sewn in Eastern Europe. The major logistical question was: how were 25,000 employees to receive new uniforms that actually fit? Uta Kötting, [responsible for this 'new corporate clothing project' at Lufthansa says]: 'If we had simply sent these items to our employees, there would have been many individual requests for alterations.' That is why a lounge formerly used for travel groups was converted to a gigantic fitting room in which thousands of items from the new uniform line, including all the odd sizes, were laid out. Uniforms were fitted piece by piece. Most of the employees were literally hustled through the process, carrying their data from one fitting station to the next."[7]

"Ultimately, about 600,000 meters of the blue fabric developed especially for Lufthansa was tailored to make up 490,000 uniform items (skirts, trousers, blazers, vests etc.)."[8]

And in the same year Lufthansa's new style was assessed in a general survey of current trends for flight attendants' fashions in the *Frankfurter Allgemeine Sonntagszeitung*:

"While the turquoise and canary yellow outfits of the seventies didn't leave much 'breathing room' for the legs and the eighties overreacted with inverted pleats in dark blue, designer Gabriele Strehle has blended restraint with sex appeal in the crew uniforms launched in May: padded shoulders, figure-hugging lines and knee-length skirts demonstrate vivacity and self-confidence. The pillbox hats are a coquettish retro element. The soaring-crane airline pays homage to pop culture by reinforcing the presence of its logo with three distinctive, yellow-gold stripes which – now for female personnel, too – designate rank."[9]

Following a system almost like the one used by national armed forces, Lufthansa maintains a service center for uniforms – the so-called "Pillbox" – at the Frankfurt operations base. The clothing needs of the flight and cabin crews are handled here, just as are the requisitions submitted by the customer service staff and ground crews. The service center for uniforms is available to all employees who are required to wear corporate clothing and offers personal fittings and advice. As a representative example of the diversity and variety of needs in uniforms, which also of course include those for areas normally hidden from passenger view, we might look at the new ramp agent's uniform, which combines remarkable colors with defined outlines. This service center also attempts, wherever possible, to archive the outfits and accessories introduced over the decades. These include shoes and handbags, scarves and hats and the unfathomable array of pins and identification plates, along with cap trim, insignia of rank, epaulettes and name plates for the various functional groups.

The insignia for the flight and cabin crews range from the captain's cap and the jacket pins worn by the pilots, flight attendants, stewards, pursers and flight engineers, through to stripes on the uniform indicating rank. They express a hierarchic attitude which is apparently rooted in the military. Even the otherwise unadorned nameplates use differing typefaces for men and women. This need for differentiation among the individual functional groups subjectively flatters the ego. Objectively, it serves for more exact identification of employees by the customers – provided that they are able to decipher the complicated codes of the hierarchic insignia. Studies conducted for nameplates bearing the crane emblem flanked by a wing on either side showed a spread in the functional groups. Used here are barcode-like stripes which are almost heraldic in character. The drafts for braids and stripes for the uniform sleeves exemplify this need for differentiation. In civil aviation the rigid pecking order has, in contrast to the military, long since given way to simple representative symbols and thus has been

Strenesse: Uniform Flugbegleiterin, ab 2002
Strenesse: Flight attendants' uniform, as of 2002

**1** Ingrid Loschek, *Reclams Mode- und Kostümlexikon*, Stuttgart 1999, S. 522.
**2** Zit. nach „50 Jahre Mode über den Wolken", in: Lufthansa.newslink spezial.lifestyle, 9/2004; http://presse.lufthansa.com.
**3** Zit. nach newslink, a. a. O.
**4** Zit. nach newslink, a. a. O.
**5** Zit. nach newslink, a. a. O.
**6** Juliane Grützner und W. O. Geberzahn, „Mode, Mysterium und Design. Gabriele Strehle formt das Label Strenesse zu einer internationalen Marke", *Design Report*, Nr. 10, 2001, S. 52–55.
**7** Petra Mies, „Mode zum Abheben", *Frankfurter Rundschau Magazin*, 5.1.2002.
**8** Zit. nach newslink, a. a. O.
**9** Ingeborg Harms, „Farben des Himmels. Pop-Kraniche, Orient-Formel und Hedonisten-Schick: Die Mode der Flugbegleiter. Eine Stilkritik", *Frankfurter Allgemeine Sonntagszeitung*, 30.6.2002, S. V2, V3.

Ausstattungswünsche der Cockpit- und Kabinencrews ebenso bedient wie jene des Boden- und des technischen Personals. Das Servicecenter Dienstbekleidung steht allen dienstbekleidungspflichtigen Mitarbeitern für eine persönliche Anprobe sowie Beratung zur Verfügung. Als stellvertretendes Beispiel für die Vielfältigkeit und Variantenbreite des Bedarfs an Dienstkleidung, der natürlich auch Arbeitsfelder umgreift, die nicht mit Passagieren in Berührung kommen, mag die neue aktuelle Ramp-Agent-Dienstbekleidung stehen, die eine bemerkenswerte Farbigkeit mit präzisen Konturen kombiniert. In diesem Servicecenter wird aber auch versucht, wenigstens überwiegend, die Outfits und Accessoires über die Jahrzehnte hinweg – wie Schuhe und Handtaschen, Schals und Hüte sowie Stoffproben und das unübersehbare Arsenal von Ansteckern und Identifizierungsschildern wie Mützenabzeichen, Rangabzeichen, Schulterschlaufen und Namensschilder für die verschiedenen Funktionsgruppen – zu archivieren.

In den Rangabzeichen für das fliegende Personal, von den Kapitänsmützen über die Ansteckabzeichen für die Jacken und Jacketts der Kapitäne, Stewards, Purser, Flugingenieure, Stewardessen und Purseretten bis zu rangdifferenzierenden Streifen auf den Uniformen kommt nach wie vor ein hierarchisches Denken zum Ausdruck, welches seine Wurzeln offensichtlich in militärischen Erinnerungen hat. Selbst reine Namensschilder verwenden für Frauen und Männer unterschiedliche Schriftarten. Dieses Differenzierungsbedürfnis der einzelnen Funktionsgruppen untereinander schmeichelt subjektiv dem Ego und dient objektiv einer genaueren Identifizierbarkeit der Mitarbeiter durch die Kunden, sofern sie die komplizierten Codes der hierarchisierenden Kennzeichnung überhaupt entschlüsseln können. Studien für Namenschilder mit dem jeweils von zwei Schwingen flankierten Kranichlogo zeigen denn auch eine Aufspreizung der Funktionsgruppen, die mit fast schon heraldischen Strichcodes operiert. Und auch die Entwürfe für Litzen und Stressen für die Ärmelkennzeichnung der Uniformen zeigen dieses Differenzierungsbedürfnis. Nun ist längst die rigide Hackordnung bei den zivilen Luftfahrtgesellschaften im Gegensatz zum Militär in Richtung bloßer Repräsentationssymbole humanisiert. Es sind nur noch graphisch-bildliche Insignien für Funktion und Orientierung, sicherlich auch Hierarchie und Einfluss, die da kultiviert werden. Aber solche Symbole haben Kraft: Sie sind ein konzises Zeichensystem, stehen für Ordnung, Struktur und Normen. Insofern vermitteln solche Rangabzeichen Werte wie Solidität und Zuverlässigkeit, Tradition und Kontinuität. Sie erinnern uns, die Passagiere und Kunden, an Zeiten, in denen, wie der konservative Schweizer Kulturphilosoph Arnold Gehlen einmal formulierte, „noch durchgegriffener gelebt" wurde. Das ist zweifellos nostalgisch, hat und behält aber seine Orientierungsfunktion.

Aber wir wollen dieses Kapitel nicht bloß archivalisch ausklingen lassen, denn Mode hat ja immer mit Emotion, Erotik und Ausstrahlung zu tun. Die jeweils neuen Uniformlinien werden vor ihrem Einsatz natürlich ebenso wie die anderen Komponenten eines Erscheinungsbildes dem Vorstand und den Mitarbeitern zur Begutachtung vorgestellt. Und da greift wohl immer auch ein tief verwurzeltes „Firmenbewusstsein". Nichts ist ja angeblich so flüchtig und so schnell „Schnee von gestern" wie die Mode. Da zeichnet im Verhältnis zu jährlich wechselnden Haute-Couture-Präsentationen die etwa alle 5–8 Jahre wechselnden Kollektionen der Airline-Uniformen schon fast eine beharrende Kontinuität aus. Neben ökonomischen Erwägungen sind dafür auch Corporate-Culture-Gründe ausschlaggebend. Ein neues Erscheinungsbild findet ja seine Akzeptanz nicht im Augenblick seiner Implantierung, sondern auf Grund von Gewöhnungsprozessen der wahrnehmenden Kunden. Die Mode der Flugbegleiter und des Personals der Fluglinien überhaupt bewegt sich insofern immer zwischen Beharrung und Innovation.

Kapitänsmütze, bis 2001
Captain's cap, through 2001

Rangabzeichen für Kapitän, Copilot, Purser und Flugbegleiter, bis 2001
Insignia of rank for captain, co-pilot, purser and flight attendant, through 2001

humanized. Only graphic and visual insignia indicating functions and orientation, but certainly hierarchy and influence, as well, are cultivated today. Even symbols like these have power. They are a concise system of symbols, standing for order, structure and standards. To this extent these insignia of rank project values such as steadfastness and reliability, tradition and continuity. They remind us, passengers and customers, of times in which, as the conservative Swiss cultural philosopher Arnold Gehlen once said, "life was lived in narrower structures." This is undoubtedly nostalgic, but does have and retains its orientation function.

But we don't want to close this chapter with a mere tour through the archives, because fashions always have something to do with emotion, eroticism and charm. It goes without saying that the new lines of uniforms, just like other components in the corporate identity, are presented to the Executive Board and employees for their evaluation before they are adopted for use. And new fashions often collide with a deeply rooted "corporate awareness." Supposedly, nothing is so volatile and so quickly out of date as fashions. When compared to the annual changes in the *haute couture* presentations, the collections for the airline uniforms, replaced every five to eight years, demonstrate almost steadfast continuity. In addition to cost considerations, reasons found in the corporate culture are also decisive. A newly introduced aspect in corporate identity will never meet with full acceptance as soon as it is implemented but, instead, as the customers who perceive the change become accustomed to it. Consequently fashions for flight attendants and all other airline employees always seek to strike a balance between continuity and innovation.

**1** Ingrid Loschek, *Reclams Mode- und Kostümlexikon*, Stuttgart, 1999, p. 522.
**2** Quoted from "50 Jahre Mode über den Wolken," in: Lufthansa.newslink spezial.lifestyle, 9/2004; http://presse.lufthansa.com.
**3** Quoted from newslink, op. cit.
**4** Quoted from newslink, op. cit.
**5** Quoted from newslink, op. cit.
**6** Juliane Grützner and W. O. Geberzahn, "Mode, Mysterium und Design. Gabriele Strehle formt das Label Strenesse zu einer internationalen Marke," *Design Report*, no. 10, 2001, pp. 52–55.
**7** Petra Mies, "Mode zum Abheben," *Frankfurter Rundschau Magazin*, January 5, 2002.
**8** Quoted from newslink, op. cit.
**9** Ingeborg Harms, "Farben des Himmels. Pop-Kraniche, Orient-Formel und Hedonisten-Schick: Die Mode der Flugbegleiter. Eine Stilkritik," in: *Frankfurter Allgemeine Sonntagszeitung*, June 30, 2002 pp. V2, V3.

Strenesse: Uniformen Flugbegleiterinnen, ab 2002
Strenesse: Flight attendants' uniforms, as of 2002

Uli Richter: 1955–1965
Uli Richter: 1955–1965

Uli Richter: Senator-Dress, 1961–1965
Uli Richter: Senator dress, 1961–1965

Uli Richter: Tropenkostüm, 1961–1965
Uli Richter: Tropical suit, 1961–1965

Ursula Tautz: 1965–1970
Ursula Tautz: 1965–1970

Ursula Tautz: Tropenkostüm, 1966–1970
Ursula Tautz: Tropical suit, 1966–1970

Werner Machnik: 1970–1979
Werner Machnik: 1970–1978

Werner Machnik: Tropenkostüm, 1970–1979
Werner Machnik: Tropical suit, 1970–1979

Jobis: 1979–1987
Jobis: 1979–1987

Jürgen Weiss: 1988–2001
Jürgen Weiss: 1988–2001

Jürgen Weiss: 1988–2001
Jürgen Weiss: 1988–2001

Strenesse: ab 2002
Strenesse: as of 2002

Strenesse: Uniform Copilotin und Flugkapitän, ab 2002
Strenesse: Uniforms for co-pilot and captain, as of 2002

Lufthansa
WorldShop
www.lh-worldshop.com
Travel
Trends
Taste

# Ein Stück Fliegen für Zuhause

Bordverkaufskatalog, 1970er Jahre
Inflight-sales catalog, 1970s

Neben allen anderen Möglichkeiten identitätsstiftender Maßnahmen für die Fluggäste – von der Werbung über die Bordausstattung bis zur Professionalität des Personals – gehört auch der Erwerb eines Erinnerungsstückes an die benutzte oder gewünschte Airline zum gern genutzten und immer mehr ausgebauten Arsenal der Kundenbindung. Sowohl in den Maschinen – unterschiedlich nach den Flugdistanzen sortiert, was auch zolltechnische Gründe hat – als auch als boomendes Mailunternehmen werden die unterschiedlichen Produkte zum Erwerb angepriesen. Umfangreiche Kataloge – sowohl für „Inflight Shopping" wie als Bestellkataloge am Boden – begleiten diese Maßnahmen, in denen diese Waren angeboten und über das „Miles & More"-Programm auch als Bonus ausgeschüttet werden. Dabei werden die angesammelten Bonusmeilen mit dem Geldwert der Waren korreliert. Aber diese Bonusmeilen werden nicht nur für Flüge gutgeschrieben, sondern z. B. auch bei bestimmten Hotel- oder Mietwagenbuchungen. Ursprünglich gab es einen „Inflight Shopping"-Katalog an Bord, einen „SkyShop"-Katalog am Boden und teilweise an Bord sowie einen „Miles & More"-Katalog, der am Boden zur Anwendung kam und auch an die „Miles & More"-Kunden zum Bestellen versendet wurde. Der „SkyShop"-Katalog war der einzige mit Lufthansa-spezifischen Produkten.

Inzwischen sind diese drei Katalogsorten im „Lufthansa Worldshop" zusammengefasst, der „Trend"-Produkte, „Taste"-Produkte und „Travel"-Produkte anbietet. Es geht also nicht nur um Waren, sondern auch um Dienstleistungen. Dazu gehören Erlebnisreisen wie die Teilnahme an Flugsimulatorprogrammen im LH Flight Training Center in Frankfurt am Main, Berlin oder Bremen; aber etwa auch bestellbare Video-Kassetten, die Langstreckenflüge vom Start bis zur Landung als Cockpit-Erlebnis nachvollziehbar machen, etwa nach Rio, Hongkong oder Chicago. Ebenso können Bonus Meilen z. B. gegen Anteile an Investmentfonds getauscht werden oder in bestimmten Geschäften, wie z. B. dem Modehaus Peek & Cloppenburg, gegen Markenartikel eingelöst werden.

Nun ist die Produktpalette der angebotenen Waren in den Flugzeugen und via Versandkatalog wahrlich immens: Kleidung, Sportartikel, Reisegepäcktaschen und -koffer, Schmuck, Uhren, Spirituosen, Parfüms, Konfekt, Zigaretten, Memorabilien. In den Maschinen – das versteht sich – werden natürlich nur solche Güter angeboten, die im Handgepäck transportabel sind: Lenkdrachen, Fahrräder und Buggys bleiben der

Seite 160: Lufthansa WorldShop Outlet in Neu-Isenburg bei Frankfurt, seit 2004
Page 160: Lufthansa WorldShop Outlet in Neu-Isenburg near Frankfurt, opened in 2004

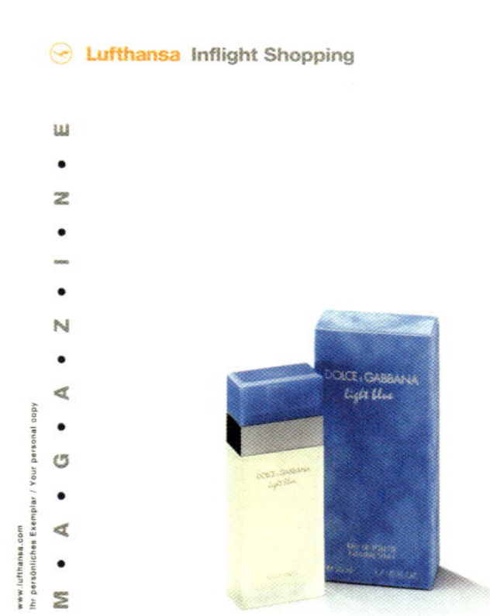


Katalog Inflight Shopping, bis 2003
Inflight Shopping catalog, through 2003

Katalog SkyShop, 2001
SkyShop catalog, 2001

Miles & More Prämienkatalog, 2002
Miles & More premiums catalog, 2002

# A bit of aviation at home

There are many techniques which can be used to instill in passengers a feeling for the corporate identity. These include creative advertising campaigns, appealing cabin interiors and the professionalism of the personnel. One method which is becoming ever more popular and is being used ever more frequently to promote customer loyalty is to provide souvenirs of the airline with which the passenger flew – or would like to have flown. Many different products are offered for purchase both in the air – the choice varying according to the distances covered and depending on customs regulations – and in the booming direct mail market. Thick catalogs – both for "inflight shopping" and for ordering on the ground – flank these campaigns. These goods are offered for sale and may also be selected as premiums in the "Miles & More" program. In this case the miles accumulated are converted into a monetary value for the products. Bonus miles can be redeemed not only for flights but also, for example, for certain hotel or rental car reservations. Originally there was an "Inflight Shopping" catalog on board, a "SkyShop" catalog for use on the ground and in some cases inflight, and a "Miles & More" catalog which again was used on the ground and was sent to Miles & More customers for their orders. The "SkyShop" catalog was the only one which featured products specific to Lufthansa.

Today these three types of catalogs have been drawn together in the Lufthansa WorldShop – the trendy products, the tasteful products and the travel products. The publication offers not only goods, but services, as well. Included here are experience-filled adventures such as taking part in a flight simulator program at one of the Lufthansa flight training centers in Frankfurt, Berlin or Bremen. Users can also order video cassettes which show every detail of long-haul flights from take-off to landing – to Rio, Hong Kong or Chicago, for instance – and experience the thrill of the cockpit. It is also possible to swap bonus miles for shares in mutual funds, for example, or to cash them in on brand items in certain businesses such as the Peek & Cloppenburg clothing stores.

The range of products on offer in the aircraft and by mail-order catalog is truly immense: clothing, sporting goods, travel bags and baggage, jewelry, watches, spirits, perfumes, sweets, cigarettes, memorabilia. It goes without saying that the items offered

Katalog SkyShop 1994/95
SkyShop catalog, 1994/95

Katalog Miles & More and SkyShop, 2002/2003
Miles & More and SkyShop catalog, 2002/2003

Katalog Inflight Shopping and WorldShop, 2004
Inflight Shopping catalog and WorldShop, 2004

Katalog WorldShop, 2004/05
WorldShop catalog, 2004/05

Lufthansa WorldShop, Outlet-Laden in Neu-Isenburg, seit 2004
Lufthansa WorldShop, outlet shop in Neu-Isenburg, opened in 2004

Katalogbestellung oder dem Direktverkauf am Boden vorbehalten, nicht nur in den Flughäfen. So hat jüngst ein Lufthansa-Shop in der Frankfurter Innenstadt im Bekleidungshaus Peek & Cloppenburg sowie ein weiterer in der Hamburger Innenstadt bei Anson' eröffnet und in Neu-Isenburg bei Frankfurt am Main gibt es sogar einen Lufthansa WorldShop-Outlet-Laden. Beileibe nicht alle Produkte dienen im engeren Sinne der Corporate Identity der Airline, es sei denn sehr indirekt. Eine Stange Marlboro-Zigaretten, einen Dimple Whisky, eine Gucci-Armbanduhr oder eine Krawatte von Hermès kann man schließlich überall kaufen: der Imagefaktor für die Fluglinie ist insofern eher indirekt und beruht darauf, dass freundlicherweise selbst solche „Fremdprodukte" als zuvorkommender Service angeboten werden. „Shopping" ist längst zu einer Kulturleistung per se geworden: Der Kaufakt als solcher bindet Identitäten. Viele, sicherlich die meisten Menschen in den industrialisierten Ländern definieren sich in ihren Statements gegenüber der Außenwelt in ihren Ingroup- und Outgroup-Verhalten auch über „Shopping". „I shop, therefore I am", statuierte die amerikanische Konzeptkünstlerin Barbara Kruger und reflektiert pointiert damit einen kulturellen Diskurs, der heute allgemeiner Konsens geworden ist – so auch für Fluggesellschaften. Im folgenden liegt die Konzentration auf jenen Gütern, oft Lizenzprodukten, die direkt und unmittelbar, über Farben, Logos und Gegenstände, den „spirit" der Lufthansa transportieren.

Die naheliegendste Produktgruppe sind kaufbare Miniaturisierungen des Fliegens selbst, also Modellflugzeuge, die bei dem Lufthansa WorldShop bestellbar sind und nach Hause geliefert werden. Wie ein früher SkyShop-Katalog vermerkt: „Keine Modellserie wurde jemals so authentisch gefertigt ... Das Modell des Aibus A300-600 erforderte allein über 2.000 Arbeitsstunden für die Konstruktion und weitere 6.000 Stunden für den Formenbau." In aller Regel werden diese Modelle in limitierter Auflage gefertigt – Mitte der 1990er Jahre im Maßstab 1:200 und für ca. 200,– DM. Damals wurden auch dazugehörige Acrylglasständer angeboten. Die Shopping-Kataloge bieten neben der legendären Wellblech-Ju 52 in ihrer „Modell-Edition" ca. weitere 20 Flugzeuge für die „Lufthansa Flotte zu Hause" an: z. B. die Boeing 747, die Boeing 747-400 mit der charakteristischen „Buckelnase" der zweigeschossigen Interkontinentalmaschinen und den City Jet B 727-200, den Airbus A340 und den Airbus A380, aber auch den 1950er-Star Lockheed L-1049, die legendäre „Super Conny". Daneben gibt es VW-, Mercedes- und Smart-Miniaturfahrzeuge mit Lufthansa-Beschriftung, die auf dem Rollfeld eingesetzt werden. Alle Modelle sind selbststehend und maßstabsgetreu (in den Maßstäben 1:500, 1:250, 1:200 bzw. 1:160) gestaltet und in Kooperation mit renommierten Modellfahrzeugherstellern realisiert. Selbst die ursprünglich für Reisebüros entwickelten großen Modelle, etwa der Boeing-Jet 747-400 im Maßstab 1:100, finden sich in den frühen SkyShop-Katalogen der Mitte der 1990er Jahre.

Zur leicht nostalgischen Anmutung der Wellblech-Junkers-Maschine passen jene von Rimowa gefertigten Alu-Bord-, Akten- und Beautycase-Koffer, die es auch als spezielle LH-Edition mit eingearbeitetem Kranich-Logo gibt: Viele Modelle, nicht alle, sind handgepäcktauglich. Ein spezieller Aluminium-Bordcase verfügt über ein großes Klarsicht-Display im Deckel, in das Austauschposter einzulegen sind. Überhaupt sind „Travel & Luggage"-Produkte naturgemäß ein naheliegendes Produktfeld. So zeigt das Angebot der LH auch Hartschalen-Trolleys aus Polycarbonat von Xenon oder von Rimowa bzw. im nostalgischen „Rillendekor" auch in Gelb, Blau oder Anthrazit eine weichere Taschenkollektion aus Nylon für sportive Aktivitäten. Zu ihr gehören auch Rucksäcke sowie Strandtaschen, ebenfalls aus Nylon, die mit LH-Strandtüchern und -Bademänteln komplettiert werden können. Im aktuellen Retro-Stil gibt es Bowling Bags, Shopper, Duffle Bags und Handtaschen aus Mikrofaserstoff, die mit dem ironischen

Lufthansa Flight Collection: Bordtasche und Trolley, ab ca. 2002, Hersteller: Titan
Lufthansa Flight Collection: Carry-on case and trolley, 2002, manufacturer: Titan

on board are only those which will fit in carry-on luggage. Kites, bicycles and buggies are reserved for catalog orders or direct sales on the ground, and not just in the airports. Recently a Lufthansa Shop was opened at the Peek & Cloppenburg clothing store in downtown Frankfurt and a further one was opened at Ansons in downtown Hamburg. There's even a Lufthansa WorldShop outlet shop in Neu-Isenburg, near Frankfurt. Certainly not all these products support the airline's corporate identity in the narrower sense; any such effect is very indirect. A carton of Marlboros, a bottle of Dimple, a Gucci wristwatch or a tie by Hermès can be purchased anywhere. The image factor for the airline is thus more indirect and is based on the fact that even such outside products are offered as a convenience to travelers. Shopping has long since become a cultural activity in and of itself. The simple act of purchasing forges identities and loyalties. Many people – and surely most of those in the industrialized countries – use shopping as one way to define themselves and their standing in regard to the "in-crowd." "I shop, therefore I am," as American concept artist Barbara Kruger once put it, thus bringing to a fine point a cultural discourse which today has become general consensus. And this is also true for air carriers. The discussion which follows concentrates on those goods, often products manufactured under license, which both directly and indirectly transport the Lufthansa spirit by way of colors, logos and objects.

The most obvious product group includes miniatures of flying itself – model airplanes which can be ordered from Lufthansa WorldShop for delivery to the home. As one earlier SkyShop catalog noted: "No models have ever been so authentic … the model for the Airbus A300-600 took 2,000 hours of work for the design alone and a further 6,000 hours for mold-making." As a rule these replicas are manufactured in limited editions. In the mid-1990s a 1:200 scale was used and the models sold for about 200 DM; matching acrylic stands were also available. The shopping catalogs feature in their "Model Edition" series not only the legendary JU 52 with its corrugated metal look but also some twenty other aircraft for the armchair traveler's "Lufthansa fleet at home." Included here are the Boeing 747, the Boeing 747-400 with the characteristic "hump" enclosing the upper deck on the intercontinental airframes, the B 727-200 City Jet, the Airbus A 340 and the Airbus 380 and, of course, the star of the fifties, the Lockheed L-1049, the legendary "Super Connie." In addition there are Volkswagen, Mercedes and Smart miniature cars sporting the Lufthansa decor as seen on the runway. All the models are, of course, exactly true to scale (at 1:500, 1:250, 1:200 or 1:160 scale), and all were realized in cooperation with renowned model makers. Even the large-scale replicas, originally developed for use in travel-agency offices – such as the Boeing 747-400 jetliner at 1:100 scale – were found in the early SkyShop catalogs published in the middle of the 1990s.

Matching the slightly nostalgic feel of the Junkers "corrugated sheet metal" airframes are the aluminum carry-on bag, briefcase and cosmetics case made by Rimowa. These are also available as a special Lufthansa edition featuring the crane logo. There are many models, not all of them in carry-on size. One special aluminum carry-on case has a large transparent display area in the lid to accept posters which may be changed out as desired. Travel and luggage products are of course quite an obvious product sector. Thus the Lufthansa range includes hard-sided trolleys made by Xenon or Rimowa either from polycarbonate or in a "corrugated" look. There's also a soft-sided luggage collection made of yellow, blue or charcoal gray nylon and suitable for more athletic activities. Included here are backpacks and beach bags, also made of nylon, which could be packed with Lufthansa beach towels and bathrobes. Available in the fashionable retro style are bowling bags, shopping bags, duffel bags and handbags

Modell Boeing 707-300, 1:500, Hersteller: Herpa
Model of the Boeing 707-300, 1:500 scale, manufacturer: Herpa

Slogan: „The swinging sixties are bag!" lanciert werden und mit einer Fotografie von Jayne Mansfield den Charme der Zeit zurückholen wollen. Ab 2004 werden diese „Retroline Bags" auch in Leder angeboten. Ebenso kultig sind Baumwolltaschen, die mit den international gebräuchlichen Flughafenkürzeln bedruckt sind. Am Image der Piloten partizipieren der Original-Lufthansa-Pilotenkoffer ebenso wie die Pilotenjacken

Bordkoffer Aluminium, ab 1998, Hersteller: Rimowa
Aluminum carry-on case, 1998, manufacturer: Rimowa

Retroline Bags, 2003 (Werbemotiv mit Jayne Mansfield, 1961)
Retroline bags, 2003 (advertising motif with Jayne Mansfield, 1961)

made of microfiber materials, lettered with the slightly ironic slogan, "The swinging sixties are bag!" and graced with a photo of Jayne Mansfield, recalling the charm of days past. These retroline bags have also been available in leather since 2004. Just as likely to attain cult status are cloth bags imprinted with three-letter international airport codes. The original Lufthansa pilot cases draw upon the aviator's image, just as do the nappa leather pilot's jackets for adults and youthful fliers. Also available is a frequent traveler wallet with pockets for various currencies and separate compartments for additional SIM cards. These are travel and holiday accessories that are positioned for use prior to, during and after the flight itself.

To be found among the travel utensils are also pocket lighters, razors, travel alarm clocks, travel hair dryers, bottle openers, sleep masks and pocket calculators. It was for these products that Lufthansa launched its first merchandising cooperation with another company – with the internationally renowned Braun AG. This choice was made due partly to its product style, reflecting the influence of the College of Design in Ulm. All these products bear the "Lufthansa" and "Braun" logotypes, both of which came from Otl Aicher's hand. The products made and marketed by Braun and by Lufthansa are highly compatible, being characterized by precision, dependability, clarity of line, rational esthetics and a common design vernacular. In the mid-1990s the first toiletry set was included in the program, more or less at short notice. It offered a safety razor, comb, shoehorn, dental care set, soap, socks, refreshing towelettes and a sleep mask for just under 30 DM. These travel accessories were created by in-house designers, and were ideal "overnight kits" for long-distance flights – compact ditty bags with the Lufthansa name, as was mentioned, stocked with toothpaste, a toothbrush, comb, refreshing towelette, safety razor and a small terry towel, and providing space for a note pad or other utensils.

The muscle-powered vehicles too, including bicycles, unicycles, bicycle trailer to transport young children, children's scooters and push-bikes – along with the ever-present jogging suit – have been made up in the Lufthansa look under license from various renowned manufacturers. They help to popularize the airline's corporate culture on playgrounds and country roads. The Lufthansa name or logo will be present almost anywhere you find sporting activities. Kites, soccer goals, tennis rackets, golfers' umbrellas and eye shades as well as equipment bags decorated with Lufthansa emblems for skis, ski boots and tennis rackets are included in the range. These are augmented by "Aviation History Caps" embroidered with a JU 52 emblem.

The "personal product designs" represent another merchandise species. Here you will find writing utensils, wristwatches and lighters. The "LH Flight Edition" line of watches features dials reminiscent of the Lockheed Super Constellation, the parabolic Lufthansa logo of the 1950s or the Airbus logotype. In addition to the nostalgic references made by these watches, there are also clear high-tech references, one example being the "Two Times Lufthansa Edition," a wristwatch which shows two time zones. When the SkyShop program was first launched there were also high-priced aviators' chronometers made by GWC (German Watch Center) for about 1,150 DM or with dials evoking images of the sea, sand and clouds. The dials, just like the calf-leather watchstraps, in gray, yellow or blue, recalled the corporate design concept drafted at the same time by the Zintzmeyer & Lux agency. Zippo lighters with Junkers or Boeing motifs remind users of bygone days while, by contrast, a lady's ball point pen and matching atomizer for perfume make use of a futuristic design language. And of course there are also leather accessories: wallets, key cases, ditty bags, manicure sets, all made of full-grain cowhide with the Lufthansa emblem. Particularly appealing were the sterling silver tie clips

„My Bordcase", 2002/03, Hersteller: Rimowa
"My Bordcase", 2002/03, manufacturer: Rimowa

Lufthansa-Braun Artikel: Feuerzeug, Schlüsselanhänger, Taschenrechner, Sonnenbrille, Elektrorasierer, Fön, 1980er Jahre
Lufthansa-Braun items: Cigarette lighter, key ring, pocket computer, sunglasses, electric razor, hair dryer, 1980s

Lufthansa Bike XC, 2003, Hersteller: Schauff
Lufthansa XC BIKE, 2003, manufacturer: Schauff

aus Nappaleder für Erwachsene und Nachwuchsflieger sowie ein Vielflieger-Portemonnaie mit Fächern für verschiedene Währungen und Extrafächern für zusätzliche SIM-Cards. Reise- und Urlaubsutensilien also, während, vor und nach dem Flug positioniert.

Zu den Reiseutensilien gehören auch Taschenfeuerzeug, Rasierapparat, Reisewecker, Reisefön, Flaschenöffner, Schlafbrille und Taschenrechner. Für diese Produkte vereinbarte die Lufthansa ihre erste Merchandise-Kooperation mit einem anderen Unternehmen: mit dem wegen seines an die HfG Ulm angelehnten Produktstils international renommierten Unternehmen Braun AG. All diese Produkte weisen die beiden Wortmarken „Lufthansa" und „Braun" auf, die im Übrigen beide von Otl Aicher stammen. Die Produkte von Braun und jene der Lufthansa sind in hohem Maße kompatibel: Präzision, Zuverlässigkeit, Klarheit, rationale Ästhetik und Formensprache. Und Mitte der 1990er Jahre wurde das First-Toilet-Set mit Nassrasierer, Kamm, Schuhlöffel, Zahnpflege-Set, Seife, Socken, Erfrischungstuch und Schlafmaske kurzerhand für knapp 30,– DM ins Programm genommen. Diese Reiseutensilien wurden von Inhouse-Entwerfern entworfen, eben jene „Overnight-Kits" für Langstreckenflüge – kleine Kulturbeutel mit Lufthansa-Schriftzug, wie gesagt, gefüllt mit Zahnpasta, Zahnbürste, Kamm, Erfrischungstuch, Mini-Rasierer und kleinem Frotteehandtuch –, Briefpapierblöcke oder Spiele, aber ebenso eigene Reisetaschen, Schlüsseletuis oder einem Stockschirm.

Aber auch die Muskelkraft-Fahrzeuge Fahrrad, Einrad, Fahrradkinderanhänger, Jogger, Kinderlaufrad und Roller, als Lizenzprodukte diverser renommierter Hersteller im Lufthansa-Design hergestellt, helfen, die Corporate Culture der Fluggesellschaft auf Spielplätzen und Landstraßen zu popularisieren. Und auf den Sportplätzen: Auch Lenkdrachen, Fußballtor, Tennisschläger, Golfschirme und Sonnensegel sowie mit LH-Emblemen verzierte Sportgerätetaschen für Skier, Skischuhe und Tennisschläger sind im Programm. Hinzu kommen „Aviation History Caps" mit aufgesticktem Ju-52-Emblem. Eine weitere Produktgattung sind die Gegenstände des „personal design": Stifte, Armbanduhren, Feuerzeuge. Die Armbanduhr „LH Flight Edition" zeigt Zifferblätter, die sich auf die Lockheed Super Constellation, das parabolische LH-Logo der 1950er Jahre oder auf den Schriftzug eines Airbusses beziehen. Neben den nostalgischen Referenzen dieser Uhren gibt es auch deutliche High-Tech-Referenzen, so z. B. bei der Armbanduhr „Two Times Lufthansa Edition", die zwei Zeitzonen bietet. Am Beginn des Sky-Shop-Angebots gab es auch hochpreisige Flieger-Chronometer vom GWC (German Watch Center) für etwa 1.150,– DM oder mit Zifferblättern, die an Meer, Sand und Wolken erinnern; wobei die Zifferblätter ebenso wie die Kalbslederarmbänder in Grau, Gelb oder Blau sehr an die zeitgleiche Corporate-Design-Programmatik der Agentur Zintzmeyer & Lux gemahnen. Auch Zippo-Feuerzeuge mit Junkers- oder Boeing-Motiven erinnern an frühere Zeiten, während wiederum ein Damenkugelschreiber mit dazu passendem Parfümzerstäuber eine futuristische Formensprache zeigt. Und natürlich die Leder-Accessoires: Portemonnaies, Schlüsseltaschen, Kulturbeutel, Maniküre-Sets – aus vollnarbigem Rindsleder mit der Lufthansa-Bildmarke. Besonders schön waren auch die Krawattenklammern aus Sterlingsilber mit dreidimensionalen Boeings, Airbussen, der alten Tante Ju oder einer DC 10. Auch die Münzen-Sets mit den bedeutendsten LH-Flugzeugen huldigten der Silbereuphorie. Sammeln, mitnehmen, erinnern: Hier wird die mentale Überschneidung zum Kapitel „Erinnerungskultur" besonders deutlich.

Und schließlich für die lieben Kleinen: Fliegerbär und Plüschflugzeug, Reiseschlafsack, Krabbeldecke, Spielservice und Spieluhr als Häschen sowie ein Kinderrucksack und der Kranich „Lu" als Plüschfigur – kuschelige Maskottchen aus waschbarem Polyester bzw. Baumwolle. Der „Lu" ziert auch eine Kinder-Reisegepäckkollektion sowie Kinder-Armbanduhren. Zum Badespaß gibt es zudem eine aufblasbare Boeing 747.

Overnight-Kits, Kunstdruck, „Lettera"-Spiel, 1980er Jahre
Overnight kits, art print, "Lettera" game, 1980s

bearing three-dimensional Boeings, Airbuses, the good old "Auntie JU" or a DC 10. Commemorative coins showing the most important Lufthansa airframes also paid homage to the fascination of silver. Collecting, making it yours, remembering. Here the mental overlap with the chapter on "the culture of recollection" becomes particularly apparent.

And finally we would not want to forget the toddlers. There are an aviator bear and plush airplane, a travel sleeping bag, blanket, toy dinnerware and a music box in the form of a bunny rabbit along with a child's knapsack and the "Lu" crane as a plush toy figure. These are cuddly mascots made of washable cotton or polyester. The "Lu" also accents a collection of children's luggage and wristwatches. And for fun in the water there is also an inflatable Boeing 747.

Neither were the adults to be forgotten. In the mid-1990s the china crafted for First Class use by designer Wolf Karnagel made its way into the SkyShop catalogs: breakfast set, dinner set, glasses. Dining culture at the edge of the stratosphere made its final descent into travelers' homes.

A separate product line with a special logotype in Lufthansa yellow has been offered since 1992/93 as giveaways by way of special distribution vans. The motto here is "up 'n' away," with the subtitle "Just fly to it" – like bees to a honey pot. The crane emblem and Lufthansa logotype are augmented with the word "Twen," referring to that in-between age from 18 to 25. The products in this range – sweatshirts, windbreakers, T-shirts, baseball caps, small bags, travel pouches and backpacks, all in charcoal gray with yellow lettering, edging and handles – have been deliberately designed to appeal to young globetrotters. These products project a feeling of "hip" and "cool."

One cooperative effort with Ford was just as hip. The outcome was a "Lufthansa Edition" of the "Ford Ka" subcompact car, which was rolled out in 1997 and limited to 3,000 vehicles, with the crane emblem on the B-pillar between the side windows. The

Fliegerchronometer Aviator, GWC (German Watch Center), 1994/95
Aviator chronometer, GWC (German Watch Center), 1994/95

Aber auch die lieben Großen: Mitte der 1990er Jahre wurde auch das First-Class-Porzellan des Designers Wolf Karnagel in die SkyShop-Kataloge gerückt: Frühstücks-Set, Menü-Set, Gläser-Set. Die Tischkultur über den Wolken im Landeanflug auf die heimischen Bungalows!

Eine eigene Produktlinie mit spezieller Wortmarke im Lufthansa-Gelb wird seit 1992/93 als „give-aways" über spezielle Distributions-Trucks angeboten. Die Wortmarke heißt „up'n away" und formuliert als Untertitel „Just fly to it" im Sinne von „darauf fliegen". Die Wort-/Bildmarke „Kranich" plus Lufthansa-Schriftzug wird um den Begriff „Twen" ergänzt. Die Produkte dieser Linie – Sweatshirts, Windbreaker, T-Shirts, Baseball-Caps, kleine Taschen, Brustbeutel und Rucksäcke, jeweils in dunklem Anthrazit mit gelbem Schriftzug, gelben Kanten und Griffen gehalten – beziehen sich so bewusst auf eine Klientel junger Globetrotter. Die Anmutung dieser Produkte ist „hip" und „cool".

Sports Collection by Lufthansa: Inline Rucksack, 2002/03
Sports collection by Lufthansa: Inline backpack, 2002/03

Ebenso hip war eine Kooperation mit Ford, die 1997 eine auf 3.000 Fahrzeuge limitierte „Lufthansa Edition" des „Ford Ka" zur Folge hatte, mit Kranich-Logo auf dem Mittelholm zwischen Seitenscheiben und jenen grauen Ledersitzen mit Paspelkanten in Lufthansa-Gelb, die auch die Clubsessel der Business Lounge, von Frog Design gestaltet, aufwies. Auch Spanien legte eine Ford-Ka-Lufthansa-Edition auf, in leuchtendem Gelb, Orange und Grün. Ein klassischer Fall von Image-Transfer! Und so fährt die LH-Identity jenseits aller Flughafengelände auf deutschen und europäischen Straßennetzen. Wenn also offensichtlich das Merchandising der Marke „LH" entsprechend umzusetzen ist, dann belegt dies sehr deutlich die führende Stellung dieser „Kranich-Marke": Der „Ford Ka" übernahm mit dieser limitierten Edition die Werte der Lufthansa – technische Kompetenz, Zuverlässigkeit, Souveränität, Perfektion. Ein Jointventure, bei dem der Autohersteller eigentlich nur gewinnen konnte. Und aufbauend auf dem Erfolg dieser Lufthansa-Edition, baute Ford 1998 3.000 Sondermodelle des Ka mit K2, dem Hersteller von Inlineskates und Snowboards, ebenfalls primär für den deutschen Markt. Gegenwärtig ist ein Motorroller von Peugeot im Lufthansa-Design erhältlich.

Und so ist der „SkyShop", der zum „WorldShop" geworden ist, ein immer größeres Dach für Vieles: Es wäre kaum verwunderlich, wenn die Reisebürotätigkeit der Flugbuchung irgendwann einmal ununterscheidbar mit dem Erwerb von Waren, Dienstleistungen fusioniert würde; „all inclusive" im erweiterten Sinn. Insgesamt zeigt der inzwischen „WorldShop" genannte Produktkatalog des LH-Merchandisings über 500 Produkte, in schöner Mischung sowohl boden- wie flugtauglich und wahlweise bezahlbar mit Euro oder Bonusmeilen. Dabei ist die Konkurrenz hart: Es gibt inzwischen allein in Deutschland mehr als 50 Mail-Order-Unternehmen – und es gibt eBay. Aber natürlich schreibt auch dieses Merchandising schwarze Zahlen; wichtiger aber ist die Kundenbindung, der Souvenir- und Trophäencharakter. Die Translocation des Flugerlebnisses via erwerbbare Ware ist insofern nur die zeitgenössische Variante jener „Mitnahmementalität", die – von Wagenfelds Salzstreuern bis zu Karnagels Edelstahlbesteck – als toleriertes Kavaliersdelikt ein Stück „Flug" umsonst in die heimischen Gefilde retten will.

gray leather seats were trimmed with Lufthansa yellow piping, just like the Business-Lounge club chairs by Frog Design. A Ford Ka Lufthansa Edition was also made in Spain – this time in bright yellow, orange and green. This was a classic case of image transfer! And thus the Lufthansa identity is on the road, far from any airfield, on German and European highways. The premier position enjoyed by the "crane" brand becomes very apparent wherever merchandising of the Lufthansa brand is to be undertaken. The Ford Ka, in this limited edition, assumed the values embodied by Lufthansa: technical expertise, reliability, preeminence, perfection. This was a joint effort from which the carmaker could not help but profit. Building on the success enjoyed by the Lufthansa Edition, Ford built 3,000 special Ka models in 1998, this time in cooperation with K2, the manufacturer of inline skates and snowboards, once again primarily for the German market. A Peugeot scooter with Lufthansa design has recently become available.

And thus the "SkyShop" has become the "WorldShop" – an ever larger "roof" covering many departments. It would hardly be surprising if classical travel agency activity – booking airline reservations – at some time in the future were to merge indistinguishably with the purchase of merchandise and services. It would become a "one-stop" vendor in the broadest sense. In the meantime the product catalog, now known as "WorldShop" and published by Lufthansa Merchandising, has grown to include more than 500 products. It is an appealing blend of items which can be used on the ground and in the air, while payment is made in euros or bonus miles, as desired. Competition is tough, however. There are more than fifty mail order companies in Germany alone – and there's e-bay as well. The merchandising efforts are, of course, in the black but more important are customer loyalty and the souvenir and "trophy" character. The translocation of the aviation experience by way of goods available for purchase is simply the contemporary embodiment of that "take-away mentality" which – from Wagenfeld's salt shakers to Karnagel's stainless steel cutlery – was a tolerated peccadillo. It let fliers take a bit of the aviation experience, gratis, home to their own four walls.

Lu Plüschfigur, ab 1994/95
Lu stuffed animal figure, 1994/95

„up'n away" T-Shirt, 1994/95, Hersteller: Adidas
"up'n away" T-Shirt, 1994/95, manufacturer: Adidas

Boeing 747-400, 1:200, Hersteller: Herpa
Boeing 747-400, 1:200 scale, manufacturer: Herpa

Junkers Ju 52/3, 1:500, Hersteller: Herpa
Junkers Ju 52/3, 1:500 scale, manufacturer: Herpa

Lockheed L-1049 Super Constellation „Super Conny", 1:500, Hersteller: Pacmin
Lockheed L-1049 Super Constellation "Super Connie," 1:500 scale, manufacturer: Pacmin

Vickers Viscount 814, 1:500, Hersteller: Herpa
Vickers Viscount 814, 1:500 scale, manufacturer: Herpa

Boeing 737-300, 1:200, Hersteller: Herpa
Boeing 737-300, 1:200 scale, manufacturer: Herpa

Boeing 737-300, 1:500, Hersteller: Herpa
Boeing 737-300, 1:500 scale, manufacturer: Herpa

Airbus A340-600, 1:500, Hersteller: Herpa
Airbus A340-600, 1:500 scale, manufacturer: Herpa

CRJ 200 Canadair Jet, CityLine, „Little Europe" , 1:500, Hersteller: Herpa
CRJ 200 Canadair Jet, CityLine, "Little Europe" , 1:500 scale, manufacturer: Herpa

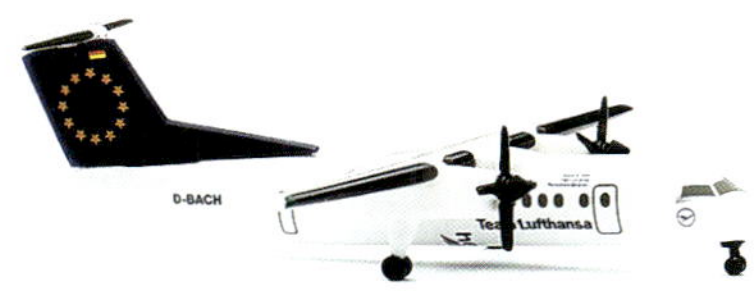

Dash 8 Q-300, „Team Lufthansa", Hersteller: Herpa
Dash 8 Q-300, "Team Lufthansa", manufacturer: Herpa

ATR 42-500, „Lufthansa Regional", 1:200, Hersteller: Herpa
ATR 42-500, "Lufthansa Regional", 1:200 scale, manufacturer: Herpa

Boeing 767-300, „Star Alliance", 1:500, Hersteller: Herpa
Boeing 767-300, "Star Alliance", 1:500 scale, manufacturer: Herpa

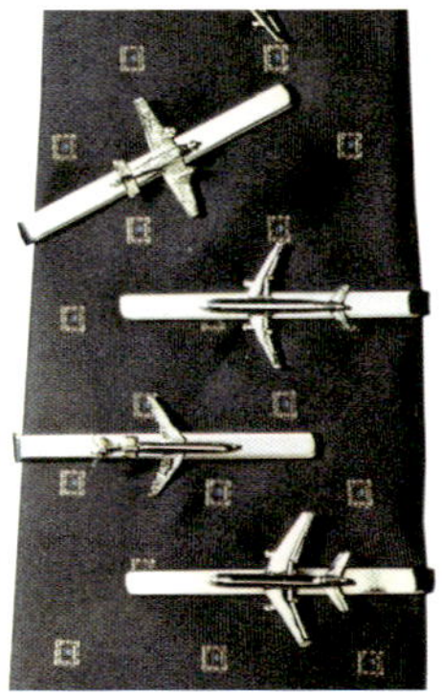

Krawattenklammern Sterling Silber, 1994/95
Tie clips, sterling sliver, 1994/95

Brieftasche, 1993/94, Hersteller: Gotta
Wallet, 1993/94, manufacturer: Gotta

Kugelschreiber mit Lufthansa-Schriftzug, Ende 1980er Jahre, Hersteller: Lamy
Ball-point pen with Lufthansa logotype, late 1980s, manufacturer: Lamy

Medaillen-Set groß, Ø 40 mm, Feinsilber, polierte Platte, Ende der 1980er Jahre
Commemorative coin set, large, 40 mm diameter, fine silver, polished plate, late 1980s

Armbanduhr „Two Times“, Lufthansa Edition, 2003
Hersteller: ITS
"Two Times" wristwatch, Lufthansa Edition, 2003, manufacturer: ITS

Kofferanhänger, Mitte 1990er Jahre
Baggage tag, mid-1990s

Schlüsselanhänger, Ende 1990er Jahre
Key ring, late 1990s

Seite 176: Anzeige für Interkontinentalflüge, um 1960
Page 176: Advertisement for intercontinental flights, ca 1960

BOEING
Jet
INTERCONTINENTAL
Der Sprung über den großen Teich ...
gelingt Ihnen im Non-Stop Flug mit dem erprobtesten Düsenverkehrsflugzeug, der LUFTHANSA Boeing Jet Intercontinental. Dieser modernste Serientyp wurde zu Ihrem Vorteil mit den besten Rolls Royce Triebwerken „Conway" und Bendix Bordradar ausgestattet. Sie werden über den ruhigen, vibrationsfreien Flug begeistert sein.
LUFTHANSA Boeing Jet Intercontinental-Dienste:
Ab 1. April um 13.30 Uhr täglich Non-Stop von Frankfurt nach New York.
Ab 12. Mai um 15.00 Uhr jeden Mittwoch und Sonnabend Non-Stop von Frankfurt nach Chicago.
Ab 13. Mai um 14.15 Uhr jeden Montag und Freitag von Frankfurt über Paris – Montreal nach San Francisco.
Direkte Anschlüsse nach Frankfurt von allen deutschen Flughäfen.
Ob Sie nun die preiswerte Economy-Klasse oder De Luxe Senator-Dienst wählen, LUFTHANSA wird Sie an Bord immer mit ihrem anerkannt guten Service überraschen. Buchen Sie jetzt schon Ihren Flug mit der zeitsparenden Boeing Jet Intercontinental. Ihr IATA-Reisebüro oder das LUFTHANSA-Stadtbüro berät Sie gern.
LUFTHANSA

# „Alles für diesen Moment“

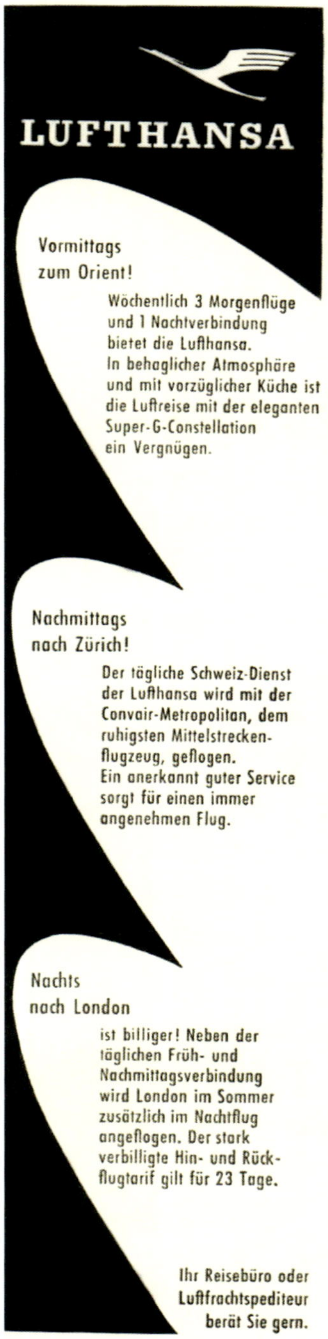


Werbung für Orient-, Schweiz- und London-Verbindungen, Ende 1950er Jahre
Advertising for routes to the Orient, Switzerland and London, late 1950s

Gute Werbung erregt Aufmerksamkeit. Um die eigenen Waren und Dienstleistungen auf dem Markt gegenüber Mitbewerbern und Konkurrenten zu positionieren, argumentiert die Werbung und preist in über die Jahrzehnte hinweg zunehmender rhetorischer Perfektion in Wort und Bild die entsprechenden Waren an. Immer aber geht es um Sympathie und Vertrauensbildung, um Psychologie und Akzeptanz sowie nicht zuletzt um gesellschaftliche Leitbilder. Werbliche Maßnahmen sollen das Selbstverständnis eines Unternehmens einem möglichst breiten Publikum vermitteln. Sie sind immer Teil einer umfassenderen Unternehmenskultur, die alle Bereiche der gestalteten Objekte – Produkte, Architektur, Mode, Gastronomie, Sprache und Bild – miteinander vernetzt. Dabei haben sich bei aller Kreativität im Einzelnen die werblichen Maßnahmen der jeweils gültigen generellen Corporate Identity einzufügen. Etwa ab Mitte der 1960er Jahre hat die Lufthansa zunehmend externe Agenturen beauftragt, die die Arbeit der hauseigenen Werbeabteilungen ergänzen und visuelle und verbale Strategien von außen einbringen. Selbstverständlich bekommen diese Agenturen jeweils ein im Unternehmen erarbeitetes „Briefing“, also eine Aufgabenbeschreibung, innerhalb derer sie sich zu bewegen haben; vergleichbar etwa den „Ausschreibungsunterlagen“ eines Architekturwettbewerbs. Dies können und sollten auch immer nur Rahmenbedingungen sein, die die persönliche Kreativität nicht behindern. Die auf solcher Basis erarbeiteten Vorschläge für Werbeanzeigen oder etwa auch für Werbespots im Fernsehen durchlaufen verschiedene, gestaffelte Entscheidungswege, ehe sie Realität werden. Da gibt es keinen Automatismus: Die Agenturen müssen den im Unternehmen für die Werbung zuständigen Abteilungen ihre Konzepte, Entwürfe und Vorschläge argumentativ erläutern, begründen und vorstellen. Deshalb ist der Weg zur Realisierung mehrstufig. Texter und Fotografen, Grafiker und Layouter stehen am Anfang: Sie erarbeiten gemeinsam den visuellen Diskurs einer Kampagne; dann kommen die sogenannten Kontakter, d. h. Kundenberater der Agenturen, die die Vorschläge im Unternehmen begründen, um dafür Zustimmung zu erhalten. Im Grunde also eine Verdoppelung: Die Werber selbst werben für ihre Werbung.

Bei jeder Werbung können die Anteile von Information und Überredung unterschiedlich gewichtet sein. Fluggesellschaften stellen in ihrer Werbung – ähnlich wie die Automobil- oder die Eisenbahnwerbung – mit wechselnden Schwerpunkten die technischen Vorzüge ihrer Beförderungsmittel, die technische und soziale Kompetenz ihrer Mitarbeiter, den Komfort ihrer Beförderung in all seinen Aspekten, die Zuverlässigkeit und Pünktlichkeit, aber auch die Umweltverträglichkeit ihres Beförderungsmittels in den Vordergrund. Bei Flugwerbung kommen die Aspekte immer entfernterer Zielflughäfen und fremder, ja exotischer Kulturen hinzu. All diese Facetten kennzeichnen auch die Werbekampagnen der Lufthansa über die letzten fünf Jahrzehnte.[1]

1957, knapp zwei Jahre nach der Wiederaufnahme des regulären Flugbetriebes, hatten die Lufthansa-Werbeanzeigen noch den Charakter von Annoncen. Sie waren grafisch äußerst einfach, zeigten den Kranich als Signet (noch ohne Kreis) und spielen abstrakt mit der Parabelform des frühen Lufthansa-Logos, das aus Kranich, Parabelschwung und dem Schriftzug „Lufthansa“ aus großspazionierten Versalien mit Serifen bestand und Dynamik, Geschwindigkeit symbolisieren sollte, aber auch an grafische Darstellungen des gesamten Globusses bzw. eines Ausschnittes von ihm erinnert. Das Deckblatt eines Lufthansa-Kalenders des gleichen Jahres, 1957, verdeutlicht diese Assoziation ebenso wie der erste Flugplan der Lufthansa, die ab 1.4.1955 „... wieder zu

# "All for this one moment"

Good advertising attracts attention. In the interest of achieving a better market position for its own products and services, in comparison to competitors, a firm's advertising presents arguments in favor of and promotes goods and services. This is done with copy and images which have become ever more rhetorically refined over the decades. In all cases, however, it is always a matter of empathy, confidence-building, psychology and acceptance. Above all, it is a matter of society's ideals. Advertising efforts are to present to the broadest possible public a company's own view of itself and its role. Advertising is always a part of a far more comprehensive corporate culture which embraces everything which the company might in any way design. That culture affects product design, architecture, fashions, meals, language and images. Here, in spite of all the creative leeway granted in specific projects, advertising efforts have to mesh with the concept for the corporate identity valid at the particular moment. Beginning in the mid-1960s Lufthansa ever more frequently commissioned external agencies to augment the work carried on by the company's own advertising departments and to contribute "outside" strategies, both visual and verbal. It goes without saying that briefings for these agencies were prepared by the company, describing both the task and the limits within which the agency was to work. This might be compared with the specifications drafted for an architecture competition. These can, and should, always be no more than a broad outline for the work, one which does not restrict personal creativity. The proposals for print advertisements or television commercials worked out on this basis pass through many stages in the decision-making processes before they become reality. And approval is by no means automatic. The agencies have to present, explain and justify their concepts, drafts and proposals to the departments responsible within the company. That is why the road to realization covers several stages. Copywriters and photographers, graphic designers and layout specialists are the project's authors. Working together, they develop the visual dialog within a campaign. Then the agency's account executives take over to explain the proposals to the client in order to gain approval. This task is actually a kind of duplication: the advertising specialists themselves are selling their own advertising concepts.

In any advertising project the weights assigned to pure information and customer persuasion may vary. Airline advertising – similar to automotive and railway advertising – sets accents which change over time. Factors which might be in the foreground are technical superiority of their aircraft or vehicles, their employees' technical and interpersonal expertise, the comfort of travelling with a particular carrier, or the reliability, punctuality and the environment-friendliness of the carrier's equipment. In air travel advertis-ing all this is augmented with the appeal of ever more distant destinations and foreign and even exotic cultures. All these facets have been used in Lufthansa's advertising campaigns over the past five decades[1].

In 1957, barely two years after regularly scheduled service was reinstated, Lufthansa advertisements still seemed more like public announcements. Extremely simple in terms of their graphic design, they displayed the crane (as yet without the circle) and engaged in some abstract experimentation with the parabolic shape found in the early Lufthansa logo. It had been composed of the crane, the energy-charged parabola and the Lufthansa name, written entirely in widely spaced capital letters with serifs. This was to symbolize dynamics and speed but also to suggest graphic depictions of the entire globe or a segment of the globe. The cover sheet for a Lufthansa calendar for the same year, 1957,

Werbung für Langstreckenflüge, Brüssel- und Rom-Verbindungen, Ende 1950er Jahre
Advertising for long-haul flights and routes to Brussels and Rome, late 1950s

Werbung „Vati fliegt Boeing“, 1960
Advertisement "Papa is flying with a Boeing", 1960

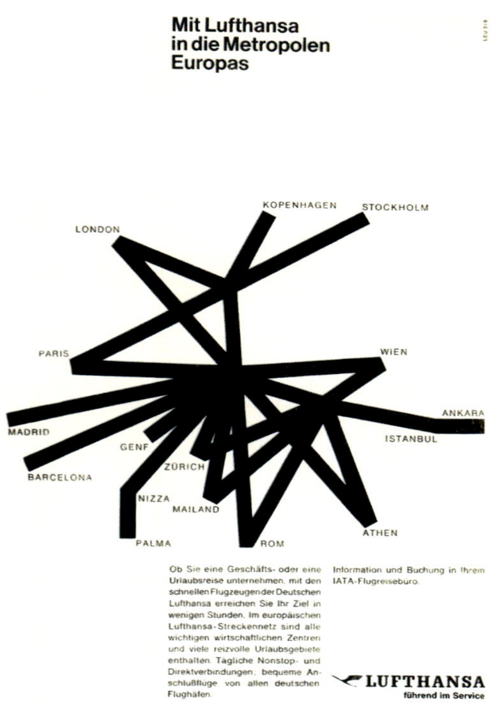


„... in die Metropolen Europas“, um 1960
"... to the European metropolises", about 1960

Ihren Diensten!“ war. Nicht zuletzt ist gestaltpsychologisch die Parabel auch das Äquivalent zur „Nase“ eines Passagierflugzeuges. Und: Der Parabelschwung gehörte auch ins ästhetische Vokabular der „Nierentischzeit“ mit den geschwungenen Bungalow-Vordächern, Tütenlampen und unregelmäßigen Bullaugenfenstern. Noch ätherischer und mit geradezu kalligrafischer Eleganz gezeichnet, wirbt ein Plakat für die Lufthansa, auf dem der Logo-Kranich selbst zum dynamisch aufsteigenden Flugzeug wird, das sich spiralförmig von der Startbahn emporschraubt. 1958 dann wird der Globus tatsächlich abgebildet, gleich zweimal, und ein agiles Strichmännchen mit Regenschirm und Sonnenhut, Aktentasche und Krawatte sowie keck blitzender Tageszeitung im Jackett visualisiert im Sauseschritt die Geschwindigkeit des Fluges über den Atlantik. Der gestrichelte Passagier ziert weitere Anzeigen und wird auch bereits mit Kind und Kegel gezeigt: Hinweis darauf, dass es schon damals nicht nur Geschäfts-, sondern auch wieder Urlaubsreisen gab.

Den Familienbezug stellt auch eine Anzeige in den Vordergrund, die 1960 erstmals ein reales Foto zeigt. Ein Kleinkind schaut sehnsüchtig auf das über ihm schwebende Flugzeuglogo und der Textblock titelt: „Vati fliegt Boeing“, und lässt im Fließtext dann verlauten: „Morgen kann er schon wieder zurück sein.“ Das hat, schon fast unfreiwillig, durchaus gewerkschaftliche Züge, forderte doch die SPD auf Wahlplakaten jener Jahre zur Einführung der Fünf-Tage-Woche auf: „Am Samstag gehört Vati uns.“ Gleichzeitig gibt es aber auch ein fotorealistisches Aquarell als Werbung für die Lufthansa, das eine mütterliche Stewardess mit einem kleinen Mädchen auf einer grazilen Flugzeugtreppe kurz vor dem Einsteigen zeigt. Das Bild strahlt jene Mischung aus Optimismus und Bodenständigkeit aus, die für die „kleinbürgerlichen“ Familienbilder des damals sehr populären amerikanischen Illustrators Norman Rockwell charakteristisch ist und die sich gerade im Nachkriegsdeutschland großer Beliebtheit erfreute: So wiesen viele Cover der Illustrierten *Hörzu* jener Jahre Illustrationen von Rockwell auf. Dessen „biedermeierlicher“ Darstellungsstil verkörperte perfekt die Sehnsucht der bürgerlichen Mittelschichten beiderseits des Atlantiks, der soeben erlebten kalten Hochtechnologie der Kriegsaufrüstung, der Waffenfabriken und der nicht zuletzt technischen Materialschlachten ein „vorindustrielles“ Gegenbild entgegenzusetzen. Jener „Sprung über den großen Teich“, wie man euphemistisch damals den Atlantik nannte, wird im gleichen Jahr mit dem freigestellten Foto des springenden, gute Laune versprühenden Geschäftsmannes visualisiert, dem offenbar auch die beiden nicht eben kleinen Reisetaschen in beiden Händen die Freude an diesem „großen Hüpfer“ nicht vergällen können. Abgestuft blasser werdende „Nachbilder“ des Springers sind hier das grafische Mittel, um Dynamik und Sprungkraft in eine optische Entsprechung zu transferieren. 1961, immer noch in Schwarzweiß, werden neue, weit entfernte Anflugziele beworben und mit abgebildeten Reisenden personalisiert: so etwa für Afrika oder Fernost. In einer Kampagne des Jahres 1963 werden dann die Personen durch jeweils landestypische Objekte ersetzt: Die Flamencogitarre steht für Südamerika, chinesisches Porzellan und Essstäbchen für Fernost, ein orientalisches Mocca-Service für Nahost und ein amerikanisches Tastentelefon der Zeit für Nordamerika. 1965 gibt es die ersten Farbanzeigen: Auch sie thematisieren die Exotik ferner Erdteile. Im gleichen Jahr, nach zehn Jahren Flugverkehr der neuen Lufthansa, leistet sie sich erstmals in einer Werbebotschaft Ironie. Da heißt es: „Können Teppiche fliegen? Ja, als Souvenir aus dem Orient – im Frachtraum des Europa-Jet.“ Das Bild der Anzeige zeigt zwei deutsche Libanon-Urlauber in einem vollgestopften Basar, die sich gerade von einem verdächtig tief verhangenen Araber eben jenen Teppich aufreden lassen. Im gleichen Jahr ziert eine Werbeanzeige für den Flug nach Johannesburg eine allerdings eher unfreiwillige Ironie, zeigt doch das abge-

made these associations clear. So did the first timetable published by Lufthansa, which went into effect on April 1, 1955, and announced: "... at your service again!" Finally, as far as design psychology goes, the parabola also suggests the nose of a passenger airplane. This parabolic figure also belongs to the esthetic vocabulary of the era of the "kidney-shaped coffee table" with flowing bungalow roofs, rice-paper lamps and irregularly placed porthole windows. Even more ethereal and drawn with calligraphic elegance, one poster promoting Lufthansa shows the crane logo itself transformed into an airplane climbing in a dynamic spiral from the runway. In 1958 the globe was then actually shown – twice, in fact. A hurrying stickman with an umbrella and Panama hat, briefcase and necktie, and a daily newspaper peeking out of his jacket represents – at a gallop – the speed of the transatlantic flight. In other advertisements this same passenger was shown with wife and children, suggesting that as early as in the late fifties people were flying not only for business purposes, but that holiday travel was also on the airline's agenda.

The reference to the family is also made in the foreground of an advertisement which, in 1960, showed a real photograph for the first time. A small child is gazing longingly at the aircraft logo floating above. The photo is captioned: "Papa is flying with a Boeing." The copy which follows then explains: "And he can be back tomorrow." Almost inadvertently, this statement is reminiscent of trade union demands of that period. Campaign posters published by the Social Democratic Party that year called for the introduction of the five-day work week: "On Saturday daddy belongs to us." At the same time, however, there is also a photo-realist watercolor used as advertising for Lufthansa, showing a mothering stewardess with a small girl on graceful air stairs just prior to boarding the plane. The picture radiates an optimistic and down-to-earth feeling characteristic of the *petit bourgeois* family images of American illustrator Norman Rockwell, whose work enjoyed great popularity particularly in post-war Germany. Rockwell's illustrations were, for example, often used on the front cover of the *Hörzu* television guide during those years. Rockwell's "homey" style embodied perfectly the longings of the middle classes on both sides of the Atlantic. It offered a comforting "pre-industrial" contrast to the recently experienced cold technology of armaments and weapons factories, not to forget the technical battles of man vs. machine. That "hop over the big pond" – as the Atlantic was then referred to – was caught on film in the same year with a photograph of a businessman literally leaping and radiating good humor. Apparently not even the two suitcases, anything but small, he carries are able to sully his enjoyment of this "great leap." Trailing visual "echoes" are the graphic effect used to lend dynamics and pep to the visual interpretation. In 1961, and still in black and white, new and far-flung destinations – such as Africa and the Far East – were promoted and personalized with pictures of travelers. In an advertising campaign in 1963 the people shown were juxtaposed with objects typical of the country mentioned: a flamenco guitar stands for South America, Chinese porcelain and chopsticks for the Far East, an oriental mocha coffee service for the Middle East and a US business telephone with its flashing buttons for North America. The first full-color advertisements appear in 1965, once again using the exotic flair of distant continents as their subject. In that same year, following ten years of scheduled service by the new Lufthansa, a touch of irony was used for the first time in a Lufthansa advertising message. The copy reads: "Can carpets really fly? Yes, as a souvenir of the Orient – in the luggage hold of the Europa Jet." The illustration shows two German vacationers in a crowded Lebanese bazaar, negotiating a price for that very carpet with a dark-complected merchant. In the same year an advertisement for the flight to Johannesburg shows a touch of involuntary irony. The

Plakat, Anfang 1960er Jahre
Poster, early 1960s

Plakatwerbung für Reisebüro, Ende 1950er Jahre
Poster for travel-agency use, late 1950s

Zeitungsanzeigen für „Super Star“ und „Super-G-Constellation“-Verbindungen, Ende 1950er Jahre
Newspaper ads for the “Super Star” and “Super-G-Constellation” routes, late 1950s

aircraft in the illustration still bears the old crane and the old logotype while the copy uses the newly introduced crane inside the circle and the new sans-serif logotype. This advertisement would thus certainly not have set well with design philosopher Otl Aicher, who was responsible for Lufthansa's overall corporate identity as of 1962. That is unless – even though this seems unlikely – the two divergent graphic depictions were deliberate, to mark the transition from one image to the other.

Beginning in about 1968 the "soft" aspects of flying were emphasized in advertising. The performance of the aircraft, technical brilliance, and pride in serving distant destinations were no longer in the spotlight. Highlighted instead were meals on board and brightly colored "pop" dinnerware, comfortable airline seating and the climate control systems in the aircraft. Leisure-oriented advertisements make their appearance in the same era. Business traveler reclines on an air mattress in a hotel pool somewhere on Jamaica, wearing sunglasses and holding a long drink. But the air carrier hastens to mention in the copy: "Layovers won't cost your company a penny." Another campaign – entitled "Lufthansa service. Part two." – promises to tend to passengers' needs on the ground, particularly during stopovers and layovers. The "second-tier service" also knows where travelers can best purchase souvenirs while showing, by way of illustration, a rather shady looking Asian offering a plethora of jewelry pinned to the lining of his jacket. But it would be too simplistic to interpret these advertisements, with their potential appeal to the "counterculture," as nothing more than a response to the social upheavals of the late sixties with their critique of the establishment, of the primacy of the modern and of the "sacred cows" of technical rationalism, instrumental reason and industrialized society as a whole. Rather, there is increased sensitivity for the "difficulties of the whole," for the diversity of the world, its peoples and their emotions. In the rhetorical view, it is with such advertisements that Lufthansa overcomes eurocentricity which, at the end of the 1960s, still prevailed in other large companies in Germany.

At the beginning of the 1970s the youth-oriented culture comes fully to the fore. New prices for young people between 12 and 25 years of age are promoted along with new destinations, such as Warsaw, which have become accessible pursuant to recent political agreements. Favorable exchange rates between the dollar and the deutschmark are the motivation for promoting air travel to America, using the photographic interpretation of a popular German phrase. Travelling to New York costs "just an apple and an egg" – next to nothing. Surrealistic elements are also used, as in 1973 with a photographic collage of an "Antigirzebrauss," a fantasy figure composed of an antelope and giraffe, zebra and ostrich. It advertises the "African adventure." In a further campaign in the mid-seventies Lufthansa employees are moved into the foreground as figures of the corporate identity: the ground staff, the experts for tariffs and tourism, the pursers and chefs, the flight engineers, the physicians and the cargo planners.

A new aircraft going into service was always the occasion for special publicity. This was the case in 1976 for the Airbus A300 B, using an interior photograph showing the seats upholstered in orange. Orange and brown were the preferred colors during those years and velvet corduroy was the favorite material. Here, again, minimalist advertisements which speakers of English might call "understated" were published in those years. Images of the sun and moon were commented upon, for example, with the line, "Two heavenly bodies which are more punctual than Lufthansa." Because of this virtue they were further characterized as "shining examples." Beginning at that same time the emblem and logotype were augmented, for about two or three years, with a sub-line which reads: "We're there when you need us." In 1977 there was a double-page spread featuring the entire fleet of aircraft, all shown to scale – the relatively small B 737 City Jet,

Service-Kampagne, um 1968
Service campaign, about 1968

bildete Flugzeug noch den alten Kranich und den alten Schriftzug, während im Text dann bereits der neu eingeführte Kranich im Kreis und der neue serifenlose Schriftzug Verwendung finden. Wohl kaum hätte diese Anzeige deshalb dem ab 1962 für die gesamte Corporate Identity der Lufthansa zuständigen Designphilosophen Otl Aicher behagt; es sei denn, obwohl dies eher unwahrscheinlich sein dürfte, dass die beiden unterschiedlichen grafischen Auftritte bewusst den Übergang vom einen zum anderen Erscheinungsbild markieren sollten.

Etwa ab 1968 werden die „soften" Leistungen des Fliegens werblich herausgestellt: Nicht mehr die Leistungsfähigkeit der Maschinen, ihre technische Brillanz, nicht mehr der Stolz auf die Erreichbarkeit ferner Erdteile, sondern das Essen an Bord und buntes Popgeschirr, der Komfort der Flugzeugsitze und die Klimaanlage im Flieger sind nun die Attraktivitätsfaktoren. In der gleichen Zeit kommen „leasure time"-Anzeigen zur Geltung. Geschäftsreisende liegen auf der Luftmatratze im Pool irgendwo auf Jamaika mit Sonnenbrille und Longdrink. Und die Fluggesellschaft beeilt sich im Text festzuhalten: „Flugunterbrechungen kosten Ihre Firma keinen Pfennig." Ebenso verspricht die Kampagne unter der Überschrift: „Lufthansa-Service. Zweiter Teil", sich gerade bei Flugunterbrechungen und Zwischenstopps um die „Bodenbedürfnisse" der Passagiere zu kümmern. Und der „Zweite-Teil-Service" weiß dann auch, wo man Souvenirs besser nicht einkauft, und zeigt zur Illustration einen dubiosen Rotlicht-Asiaten mit überquellenden Schmuckangeboten auf dem Innenfutter seines Jacketts. Aber es wäre zu einfach, solche potenziell „subkulturellen" Anzeigen einfach nur als Reflex auf die 1968er-Bewegung mit ihrer Kritik am Etablierten, am Königsweg der Moderne und an den „alten Zöpfen" der technischen Rationalität, der instrumentellen Vernunft und der Industriegesellschaften insgesamt in eins zu setzen. Vielmehr zeigt sich da eine erhöhte Sensibilität für das „schwierige Ganze", für die Vielfältigkeit der Welt, ihrer Bewohner und deren Emotionen. Rhetorisch gesagt: Mit solchen Anzeigen überwindet die Lufthansa einen Eurozentrismus, der am Ende der 1960er Jahre bei anderen nationalen Großunternehmen durchaus noch vorherrschte.

Anfang der 1970er Jahre kommt dann die Jugendkultur voll zum Tragen. Neue Jugendtarife für Teens und Twens zwischen zwölf und 25 Jahren werden beworben, aber auch neue Zielflughäfen, die aufgrund politischer Vereinbarungen möglich werden, z. B. Warschau. Günstige Wechselkurse zwischen Dollar und DM motivieren zur Bewerbung einer Flugreise nach Amerika mit dem schlagenden Bild eines Apfels und eines Hühnereis: nach New York also „für 'nen Appel und 'n Ei". Auch surreale Elemente kommen jetzt zum Tragen, etwa 1973 mit der Fotomontage eines „Antigirzebrauß", eines Fantasiewesens, zusammengesetzt aus Antilope und Giraffe, Zebra und Strauß, das für „Das Afrikanische Abenteuer" wirbt und doch irgendwie an die urdeutsche „eierlegende Wollmilchsau" erinnert. In einer weiteren Kampagne der Mittsiebzigerjahre werden dann Mitarbeiter der Lufthansa als identitätsstiftende Figuren in den Mittelpunkt gerückt: das Bodenpersonal, die Experten für Tarife und Touristik, die Purser und Küchenmeister, die Ingenieure, die Ärzte und die Frachtexperten.

Immer, wenn neue Maschinen in Dienst gestellt werden, werden sie beworben, so etwa 1976 der Airbus A300 B mit einer Innenraumaufnahme oranger Sitzreihen. Orange und Bärenbraun waren die bevorzugten Farben dieser Jahre und Cordsamt war ihr Lieblingsmaterial. Aber auch geradezu minimalistische Anzeigen, die der englische Sprachraum wohl „sophisticated" nennen würde, werden in diesen Jahren veröffentlicht. Zwei Bilder von Sonne und Mond werden z. B. kommentiert mit der Zeile: „Zwei, die pünktlicher sind als Lufthansa", und mit der Charakterisierung „Leuchtende Vorbilder" versehen. Ab nun kommt zu dem Logo und dem Schriftzug der Fluglinie für zwei, drei

Kampagne für Interkontinentalflüge, Anfang 1970er Jahre
Campaign for intercontinental flights, early 1970s

the B 737 Europa Jet, the wide-bodied Airbus A300, the long-haul B 707 and DC 10, as well as the B 747 Jumbo Jet. The advertisement is headlined: "Ornithology at Lufthansa." In 1978 the sub-line was removed and the agency then responsible showed sections of Lufthansa airframes in the advertising – details such as the landing gear, doorways or fuselage sections, almost always in conjunction with maintenance work symbolizing the care, reliability and technical expertise which the airline embodies. These detail views were joined by captions which always interpreted the image in a slightly ironic tone of voice. In regard to aircraft cleaning, for example, it was claimed that "There's no flier that's more immaculate" and "We're working on ourselves" to describe technical inspections. A crew of airfield technicians is described with the line: "Fortunately, you'll find people with a penchant for thoroughness everywhere," and an all-male crew in front of an airplane on the ground, accompanied by the sentence: "Of course, we also have hard-working stewardesses." The fascination of flying itself and of the aircraft as brilliant pieces of high-tech equipment are shown with illustrations of airplanes combined with sentences such as "Your mode of transit for nearby destinations, too" or "We're hiring: 6 new Jumbos, 3 new Airbuses, 6 new Boeing 727s." A nighttime photograph of an Airbus in flight is captioned "We keep on working, long after the shops have closed." This bit of irony lends emotional values to the technology being promoted here. This particular advertising campaign, with dozens of motifs and running through 1981, was one of the most varied campaigns Lufthansa had ever seen. Not only did it give rise to the clever statement, "It's the work on the ground that pushes an airline to new heights," but also the sentence harking back to the 1960s: "And in the evening papa belongs to the family." Once again in 1980 the results of direct, bilateral negotiations made themselves felt in Lufthansa's advertising. That year it was possible to reach political agreement so that, from that point forward, communist Beijing granted landing rights to Lufthansa flights. The airline responded with an advertisement showing exquisi-

Kampagne mit Mitarbeitern, 1975
Campaign featuring employees, 1975

Flotten-Anzeige, 1977
Advertisement showing the fleet, 1977

Jahre eine Unterzeile hinzu, die heißt: „Wir sind da, wenn Sie uns brauchen." 1977 gibt es in doppelseitigen Anzeigen die gesamte „Flugzeugflotte" in maßstabsgerechten Größenverhältnissen zu bestaunen: den relativ kleinen City-Jet B 737, den Europa-Jet B 737, den Großraumflieger Airbus A300, die Langstreckenflugzeuge B 707 und DC 10 sowie den Jumbo-Jet B 747. Die Anzeige ist übertitelt „Die Lufthansa-Vogelkunde". 1978 fällt die Logo-Unterzeile wieder weg und die nun zuständige Agentur zeigt in den Werbeanzeigen Ausschnitte von Lufthansa-Maschinen, also Details wie Fahrgestelle, Einstiegstüren oder Rumpfteile, so gut wie immer verbunden mit Wartungsarbeiten, die symbolisch für die Sorgfalt, Zuverlässigkeit und technische Kompetenz der Fluglinie stehen. Diese Ausschnittbilder werden mit Bildzeilen verbunden, die das Bild immer ironisch in Sprache übersetzen. Da heißt es dann etwa „Keiner fliegt reiner" bei Säuberungsarbeiten, „Wir arbeiten an uns" bei technischer Inspektion, oder eine Riege von Flugfeldtechnikern wird betextet mit der Zeile: „Leute mit einem Hang zur Gründlichkeit findet man glücklicherweise überall." und eine nämliche männliche Crew vor einer Maschine am Boden mit dem Satz: „Tüchtige Stewardessen haben wir natürlich auch." Die Faszination am Fliegen selbst ebenso wie an den Flugzeugen als brillante High-Tech-Maschinen zeigen abgebildete Flugzeuge, die mit Sätzen kombiniert werden wie „Ihr Nahverkehrsmittel", „Wir stellen ein: 6 neue Jumbos, 3 neue Airbusse, 5 neue Boeing 727", oder die Nachtaufnahme eines fliegenden Airbusses erhält die Zeile „Wir arbeiten auch nach Ladenschluss". Durch diese Art von Sprachironie bekommt die beworbene Technik emotionale Werte. Diese Kampagne war mit Dutzenden von Motiven bis 1981 eine der vielfältigsten Werbekampagnen der Lufthansa überhaupt und hat nicht nur die kluge Aussage „Es ist die Arbeit am Boden, die eine Fluglinie nach oben bringt" kreiert, sondern auch den auf die 1960er Jahre zurückgemünzten Satz: „Und abends gehört Papi der Familie." Auch 1980 machten sich erneut direkt zwischenstaatliche Verhandlungsergebnisse in der Lufthansa-Werbung bemerkbar. In diesem Jahr konnte politisch die Übereinkunft erzielt werden, dass von nun an auch das kommunistische Peking Lufthansa-Maschinen Landeerlaubnis erteilte. Die Fluglinie reagierte darauf mit einer Anzeige wunderschöner chinesischer kalligrafischer Schriftzeichen, die, obwohl sie in unserem Kulturraum nun sicherlich nur von einer Minderheit entziffert werden können, als grafische Signale für dieses Ereignis durchaus prägnante Wirkung hatten.

Mit ebenso dominant grafischen Mitteln wird 1982 der Austausch der City-Jets gegen die neuen Boeing 737 SL beworben: In acht Reihen wiederholen sich jeweils vier farbige Reinzeichnungen dieser 32 neuen Flugzeuge für Kurz- und Mittelstrecken. Aber generell läuft die als Serie bereits 1978 begonnene Kampagne mit den Detailbildern weiter. Zwei übereinander geschlagene, „sitzende" Herrenbeine in edlem Zwirn und vornehmem Schuhwerk stehen für erweiterten Sitzreihenabstand. Hier gibt es noch das Kranich Logo und den Schriftzug, ein paar Anzeigen später dann nur noch den Kranich, aber die Wortmarke „Lufthansa" ist nun zu dem Satz erweitert: „Der Unterschied ist Lufthansa." Bei den Folgeanzeigen fällt dann der Kranich völlig weg, z. B. bei dem Motiv eines mit Sekt mittels einer Hubbühne eingeweihten neuen Fliegers, von dem nur die Buckelnase und die Cockpit-Fenster sichtbar sind. Der Vorgang der Flugzeugtaufe wird kommentiert mit dem Satz: „1983 war ein geburtenstarker Jahrgang bei Kranichs: 6 neue Airbusse A310." Indem mit dem Satz „Der Unterschied ist Lufthansa" aus der Wortmarke ein Aussagesatz mit Behauptungscharakter wird, wird die Fluglinie nicht mehr als eine Gesellschaft beworben, die sich der Konkurrenz anderer Fluglinien zu stellen hat, sondern sie bekommt so den Charakter einer Alleinstellung gegenüber allen Mitbewerbern. Und diese „linguistische Strategie" ist doppelt klug, da der Satz semantisch abstrakt bleibt. Gemeint sein, assoziiert werden kann jeder denkbare Unter-

Service-Kampagne, 1979
Service campaign, 1979

te Chinese characters in superb calligraphy. Even though it is certain that only a small minority of those who saw it in the West would be able to interpret the meaning, it had a striking effect as a graphic signal for this event.

In 1982 the replacement of the City Jets by the new Boeing 737 SL was publicized with graphics which were just as striking. Shown there were eight rows of four colored renderings for these 32 new aircraft for short- and medium-distance flights. In general, however, the series using detail photographs and launched in 1978, continued. A pair of male legs, crossed, "sitting," and clad in fine fabric and elegant footwear, stands for increased distance between seat rows. Both the crane emblem and the logotype are still shown here. A few advertisements later one sees just the crane, but the Lufthansa logotype has been expanded to declare that: "Lufthansa makes the difference." In later advertisements the crane is omitted completely. One motif shows, for instance, a new airframe being christened with champagne with the help of an aerial platform; here only the "hump-backed nose" and the cockpit windows are visible. The ceremony is commented upon as follows: "In 1983 there was a regular baby boom among the cranes: 6 new Airbus A310s." Due to the fact that the statement "Lufthansa makes the difference" elevates the logotype to a declarative sentence, the airline is no longer publicized as a company facing competition from other airlines. Instead it assumes a unique position vis-à-vis all competitors. This "linguistic strategy" is doubly clever, since the sentence remains semantically abstract. Every conceivable difference can be read into this statement: greater friendliness on the part of the personnel, enhanced punctuality, a higher degree of technical expertise, more comfort, more luxurious dining and much, much more. In a black-and-white advertisement in 1983 the "examples" are made more concrete by way of images: a whisky on the rocks, stereo headphones, stainless steel cutlery with a damask napkin and a timetable symbolize the qualities found in Business Class. Commentaries in captions "objectivize" the emotional statements in the photographs. The impression which results is thus one of objective news reporting. An advertisement for air cargo services also emphasizes the difference to other air carriers. It shows a front view of a jumbo jet, nose raised, being loaded with large pallets of fruits and vegetables, "from every corner of the world," so that "Even in winter there'll be no shortage of vitamins." Finely honed social skills, prudence and nurturing are the subliminal messages here, while a photograph of a dark, candlelit distillery cellar, obviously centuries old, with wooden whisky casks festooned with spider webs, symbolizes patience and care. The photo is captioned: "Preparations for our First Class flights often take years." Almost surreal logical disparities between the claims in the image and in the copy, such as in this "whisky" advertisement, use initial viewer disorientation to boost the attention value considerably.

One airplane is photographed very elegantly from an elevated viewpoint and in wide-angle technology. It shimmers, noble in the night, waiting on the runway illuminated with its marker lamps and quite apparently ready for take-off. The caption: "One of the most fastidious manifestations in global air traffic." This message turns high-tech into high-touch. Business travelers, photographed on the outbound and return flights in the light of dawn and dusk, symbolize wrapping up various foreign transactions within a single day. In the morning passengers, looking toward the left, travel to the west while the evening, looking to the right, they travel eastward. The symbolism of this change of direction in the two photographs is underscored by the "activities" of the four protagonists in each case. In the morning, before all their business meetings, they have plans, files and appointment books on their laps; in the evening, quite apparently after bringing business to a successful conclusion, they relax with a drink, a lifestyle magazine or a good novel. The visual message thus formulated is so impact-laden that the lines

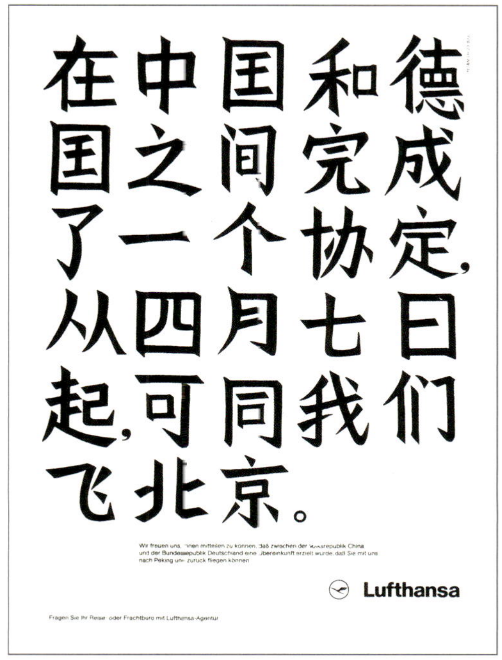


Anzeige erster Peking-Flug, 1980
Advertisement for the first flight to Beijing, 1980

Flotten-Anzeige, 1984
Advertisement showing the fleet, 1984

Service-Kampagne, 1981
Service campaign, 1981

schied: mehr Freundlichkeit des Personals, mehr Pünktlichkeit, höhere technische Kompetenz, mehr Komfort, luxuriöseres Essen und dergleichen mehr. Auf einer Schwarzweissanzeige werden 1983 dann z. B. die „Unterschiede" bildlich konkretisiert: Ein Whiskey auf Eis, ein Stereo-Kopfhörer, ein Edelstahlbesteck mit Damastserviette und ein Flugplan symbolisieren die Qualitäten der Business Class, kommentierende Bildbetextungen „versachlichen" die emotionale Aussage der Bilder. So entsteht gewissermaßen der Eindruck objektiver Nachrichten-Berichterstattung. Auf die Differenz zu anderen Fluggesellschaften hebt auch eine Anzeige für Luftfrachtbeförderung ab, deren Bildmotiv die frontal fotografierte, hochgeklappte Nase eines Jumbo-Jets ist, der von vorne mit großen Paletten voller Obst und Gemüse „aus aller Welt" beladen wird, „damit", wie es heißt, „ auch im Winter die Vitamine nicht zu kurz kommen". Soziale Kompetenz, Fürsorge, Betreuung sind hier die unterschwelligen Botschaften, während das Bild eines düsteren, offensichtlich jahrhundertealten Destilleriekellers mit spinnwebenverhangenen hölzernen Whiskeyfässern und Kerzenlicht Geduld und Sorgfalt symbolisiert. Zum Bild dann die Zeile: „Die Flugvorbereitungen unserer Ersten Klasse dauern oft Jahre." Fast surreal-logische Brüche der Aussagezusammenhänge zwischen Bild und Text wie in dieser „Whiskey"-Anzeige erhöhen durch eine anfängliche Desorientierung der angesprochenen Leser deren Aufmerksamkeitswert beträchtlich.

Ein höchst aufwendig und elegant von einem erhöhten Standpunkt aus in Weitwinkeltechnik fotografiertes Flugzeug, das bei Nacht, edel schimmernd, auf dem mit Markierungslampen erleuchteten Rollfeld offenbar gerade „ready for take-off" ist, bekommt durch die Zeile: „eine der gepflegtesten Erscheinungen im Weltluftverkehr" eine Botschaft unterlegt, die High-Tech in High-Touch ummünzt. Geschäftsreisende, beim Hin- und Rückflug im Morgenlicht bzw. in der Abenddämmerung fotografiert, visualisieren die Abwicklung diverser Auslandsgeschäfte innerhalb eines Tages: Am Morgen reisen die Passagiere, indem sie nach links schauen, gen Westen; am Abend, nach rechts schauend, gen Osten. Die Symbolik des Richtungswechsel in den beiden Bildern wird durch die „Tätigkeiten" der jeweils vier Protagonisten untermauert. Am Morgen, vor allen Geschäftstreffen, haben sie Pläne, Akten, Terminplaner auf dem Schoß; am Abend, nach offenbar erfolgreichem Geschäftsabschluss, gönnt man sich einen Drink, ein Lifestyle-Magazin oder einen guten Roman. Die so formulierte visuelle Suggestivität ist so schlagend, dass es der beigegebenen Zeile „Ob Termine in der Oxfordstreet oder an den Champs-Élysées, Sie sind zum Abendessen wieder zu Hause" kaum noch bedürfte, aber sie verstärkt die präsentative, simultane Logik der Bildaussage durch die diskursive, argumentative Sprachaussage.

1984 erfolgt zwar die Rückkehr zu Kranich und Wortmarke, aber die Agentur behält die generelle Strategie eines Bildes bei, das textlich ironisch kommentiert wird. Ein Flugbegleiter hält ein Tablett mit Whisky-Glas und vorgewärmter Babyflasche vor sich und darüber liest man: „Gerade an Kleinigkeiten erkennt man den guten Gastgeber."

Diese langjährige Kampagne war ein Meisterstück selbstevidenter, gleichwohl augenzwinkernder Souveränität, deren einzelne weit über 200 Anzeigenmotive es durchgehend fertig brachten, Schmunzeln und Amüsement auszulösen. Das postmoderne Jahrzehnt der 1980er Jahre mit seinem Hang zur Ironie, zum Laisser-faire, zur Mehrdeutigkeit, auch zum verfeinerten Luxus einer „Toskana-Fraktion" hat so auch seine Spuren im „spirit" des werblichen Auftritts der Lufthansa hinterlassen. In diesem Jahrzehnt begann auch die Inflation der doppeldeutigen Untertitel, die die jeweiligen Markennamen kommentieren und in kurzen Statements Charakterzüge des Produkts pointieren. Kein Mobiltelefon, kein Brotaufstrich, kein Automobil, keine Dienstleistung verzichtet auf diese Positionierungsmöglichkeiten.

which are appended – "No matter whether your appointment is in Oxford Street or on the Champs-Élysées, you'll be back home for dinner" – are almost superfluous. They do, however, reinforce the emotionalized, simultaneous logic of the visual statement with a supportive verbal statement.

1984 did, indeed, see the return to the crane and logotype but the agency retained the general strategy of a photograph which was commented upon with a touch of irony. A flight attendant holds a tray bearing a whisky glass and pre-heated baby bottle while above the viewer reads: "It's the little things that mark a good host."

This long-running campaign was a masterpiece of self-evident but at the same time wry authority. Each of the far more than 200 motifs were capable of triggering smiles and amusement. The post-modern decade of the 1980s with its penchant for irony, laissez-faire, ambiguity and the sophisticated luxury of the "leisure class" made its mark on the spirit embodied in Lufthansa's advertising. It was in this decade that the inflationary use of subtitles with double meanings began. These commented upon the brand names in each case and used concise statements to bring the product's characteristics to a fine point. There was hardly a mobile telephone, sandwich spread, automobile or service which eschewed this option for positioning itself in the market.

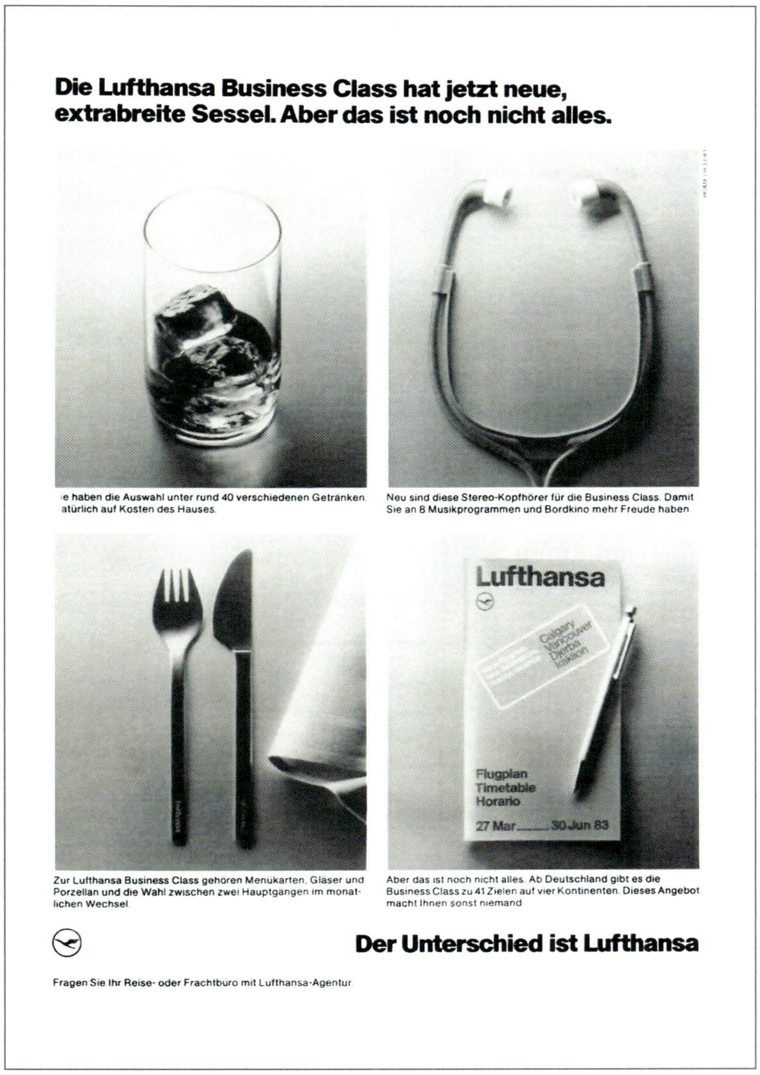


Service-Kampagne, 1983
Service campaign, 1983

Service-Kampagne, 1984
Service campaign, 1984

Service-Kampagne, 1987
Service campaign, 1987

1987 kamen abgewandelte „Lebensweisheiten“ wie „Erst die Arbeit, dann das Fliegen.“ oder „Wissen ist besser als glauben.“ hinzu, kombiniert mit den schon bekannten Ausschnittfotos von Flugzeugen am Boden mit Wartungsmonteuren. In gewisser Weise zeigen aber gerade diese „Wartungsfotos“ die Ambivalenz jeder werblichen Aussage: Denn einerseits verweisen sie auf Sorgfalt, Gründlichkeit und Zuverlässigkeit, andererseits aber auch auf die technische Fragilität und auf die Notwendigkeit, den Maschinenpark ständig zu pflegen, zu kontrollieren und zu überarbeiten. Und doch kippt dies dann wieder ins Positive um: Schließlich ist Risikominimierung eine positive Strategie und auch der Faktor „Ehrlichkeit“ ist da als positiver Imagetransfer wohl nicht zu unterschätzen. Spätestens seit Mitte der 1970er Jahre war nach dem Umweltbericht des „Club of Rome“ die Technikfolgenabschätzung als Kritik an der monokausalen Technikeuphorie auf der Tagesordnung. Indem die Lufthansawerbung diesen Diskurs aufgreift und positiv wendet, reflektiert sie durchaus auch das wachsende Bewusstsein für die Begrenztheit der Ressourcen und für die Begrenztheit aller bloß technischen Vernunft.

1988 wird dann etwa das bestellbare „Lufthansa-Sparbuch“ mit der Auflistung günstiger Tarife mit einem Sparschwein – versteckt hinter einem aufgeschlagenen Büchlein – mit der Zeile „Da wird das Sparschwein zur Leseratte“ kombiniert. Am oberen Rand der Anzeige heißt es: „Die blau-gelben Seiten des Fliegens“, womit dem Streckenplan die enzyklopädische Bandbreite eines Branchenbuches, eben jene der sogenannten „Gelben Seiten“, zugesprochen wird. Die Frischetüte mit dem sogenannten „Lufthansa Morgen Snack“ wird als „Deutschlands erfolgreichstes Pausenbrot“ vorgestellt und die Abbildung des legendären Junkers-Flugzeuges, seit 1936 von der Lufthansa eingesetzt, wird untertitelt mit: „Nur eines unserer Flugzeuge wird bei uns alt.“

Im Jahr des harten Preiskampfes um billige Flugpassagen, 1989, werden die Anzeigen strenger, fast pädagogisch: „Eine Fluggesellschaft, die moderne und zuverlässige Flugzeuge für den wichtigsten Service hält, wird nie zu den billigsten gehören. Trotzdem bietet Lufthansa mit dem ‚Nordamerika Special’ für USA und Kanada jetzt Preise, die sich im Zaum halten: zwischen DM 999.– (Ostküste) und DM 1.333.– (Westküste)“ heißt es, fast um Nachsicht bittend, unter einem marlborohaften Cowboy-Bild, das natürlich mit „Freiheit und Abenteuer“ betitelt ist. Mit ähnlich umfangreichen Textblöcken werden die deutsche Gründlichkeit und Wertarbeit sowie das Label „Made in Germany“ auf die Lufthansa bezogen. Auch 1990 setzt sich diese „Preis-Begründungsstrategie“ fort. Gleichzeitig wird die Kulturförderung der Lufthansa in den Vordergrund gerückt, etwa mit einem Engagement für eine Max-Beckmann-Ausstellung in Leipzig. Natürlich ist nach der Wiedervereinigung auch die nun erneut mögliche Präsenz in Berlin ein Thema der Zeit.

In den 1990er Jahren spielt die Lufthansa zudem die „Umweltkarte“ aus: Lärmteppich, Treibstoffverbrauch, Triebwerkkontrolle während des Fluges, Lackentfernung mit Wasser statt mit chemischen Mitteln sind nun die Argumente der Sympathiebildung. Auch der „neue Osten“ kommt ins Visier: 1992 fliegt die Lufthansa bereits 14 Großstädte Osteuropas an. 1998 werden die Werbeanzeigen geradezu esoterisch: Weich gezeichnete Blumenfotografien stehen symbolisch für die Transzendierung einer technischen Kompetenz, die nicht mehr visuell die Muskeln spielen lassen muss. Nun erscheint auch das von dem Londoner Designbüro Pentagram entwickelte Logo der Star Alliance in den Anzeigen, eines logistischen Zusammenschlusses führender, zunächst fünf internationaler Fluggesellschaften, der schnell auf die doppelte Mitgliederzahl anwächst. Keine „harten Fakten“ also mehr: Eine Doppelaufnahme von Manhattan im Schneegestöber wird lakonisch kommentiert mit „cold“ und „cool“. Dies reicht, um positiv die New-York-

Spar-Kampagne, 1988
Savings campaign, 1988

In 1987 modified aphorisms were added, such as, "Work before flying" or "Knowing is better than believing." These were combined with the familiar detail photos of aircraft and maintenance technicians on the ground. To a certain extent, however, it is just these "maintenance photographs" which demonstrate the ambivalence of the promotional claim. On the one hand, these photos point to care, thoroughness and reliability while on the other hand they emphasize the technical fragility and the necessity to groom, inspect and service the fleet. This then tips the balance once again toward the positive. Minimizing risk is ultimately a positive strategy and the "honesty factor" should not be underestimated in its power for positive image transfer. At the latest since the middle of the 1970s and following the environment report issued by the Club of Rome, the estimates of the aftermath of unbridled technology have been on the agenda – as a critique of the all-pervasive technological euphoria. Lufthansa advertising picks up this discussion and turns it to its own benefit, reflecting growing awareness of the fact that resources are finite and of the limitations in reasoning which is based solely on technology.

In 1988 the "Lufthansa savings passbook," which contained a listing of favorable tariffs, was combined with a savings bank hidden behind the open booklet, with the caption, "Here the piggy bank turns into a bookworm." At the upper edge of the advertisement it says, "The blue and yellow pages of flying,“ imparting to the timetable the encyclopedic breadth of the yellow pages. The carry bag with the so-called "Lufthansa Morning Snack" is introduced as "Germany's most successful in-between meal" and the illustration of the legendary Junkers airframe, deployed by Lufthansa since 1936, is subtitled with: "The only aircraft to reach old age in our fleet."

In 1989, the year characterized by tough price competition, advertisements became sterner and almost educational. The airline informs its clientele, almost apologetically, that:

"Any airline which sees modern and reliable airplanes as its most important service will never be among the cheapest. Despite this, Lufthansa now offers with its 'North American Special' prices for the USA and Canada which are in line with the market – between DM 999 (East Coast) and DM 1,333 (West Coast)."

The familiar-looking, Marlboro-like cowboy photograph is, of course, entitled "Freiheit und Abenteuer" – freedom and adventure. Equally concise copy is used to associate Lufthansa with German thoroughness and craftsmanship as well as the "Made in Germany" tag. This "price justification strategy" is continued in 1990. At the same time Lufthansa's sponsorship of the arts is moved into the foreground. One example is its involvement in a Max Beckmann exhibition in Leipzig. Of course, following the unification of East and West Germany, the airline's renewed presence in Berlin is a major topic of those exhilarating years.

In the 1990s Lufthansa also played the environment card: noise footprint, fuel consumption, engine monitoring during flight, and paint removal using water instead of chemicals are now the arguments used to foster empathy. The "New East" also comes into sight. By 1992 Lufthansa was already serving fourteen major cities in Eastern Europe. In 1998 advertisements became almost esoteric. Softly photographed flowers symbolize the transcendence of technical expertise which no longer has to flex its muscles visually. The logo created by the Pentagram design office in London for the Star Alliance now also makes its appearance in advertising. This is a logistical alliance among leading international airlines, initially five in number but quickly growing to twice that. No longer is advertising based on hard facts. We see two identical pictures of Manhattan during a snow storm and the laconic captions are "Cold" for the one and

Preis-Kampagne, 1989
Price-based campaign, 1989

Anzeige Wiederanflug Berlin, 1990
Advertisement announcing resumption of service to Berlin, 1990

Kulturförderung, 1990
Arts sponsorship, 1990

Streckennetzanzeige, Mitte 1990er Jahre
Route network campaign, mid-1990s

Flüge der Lufthansa zu bewerben. Jetzt lautet die Unterzeile: „You see the world the way you fly". Die damals neu eingeführten Tarifstandards der sogenannten Lufthansa Specials, Sondertarife für Geschäftsreisende, werden mit dem alten grafischen Stilmittel des Weglassens bestimmter Buchstaben eingeführt; einem Verfahren, das Boulevardblätter und Magazine seit Jahrzehnten anwenden, indem sie etwa durch überschneidende Prominentenfotos auf den Coverseiten Wortmarken wie *Stern*, *Spiegel* oder *Brigitte* nur noch rudimentär abbilden. Trotzdem ist der Wiedererkennungswert gewährleistet. Jetzt wird Gelb die Hauptfarbe für die Identifikation mit der Fluglinie: so auch bei jener Abbildung eines Ticketabschnittes für die jahrzehntelange Herausgeberin der Wochenzeitung *Die Zeit*, Marion Gräfin Dönhoff, die kommentiert wird mit dem Satz: „Sie erreicht Orte, zu denen wir sie nicht fliegen können. Die Herzen der Menschen." Die Anzeige für diese, wie einmal formuliert wurde, „Botschafterin der deutschen Befindlichkeiten in jeder Hinsicht" erschien übrigens zu ihrem 90. Geburtstag. Mit der typografischen Direktheit von Schlussverkäufen argumentiert eine Preisbrecheranzeige der „Lufthansa Special" zur Jahrtausendwende: eine Rückkehr vielleicht zu den frühen reinen Textaussagen der neuen Lufthansa in den Mittfünfzigerjahren?

Ab 2000 werden die Bildaussagen rationaler, ohne ganz auf Ironie zu verzichten. Die Beschilderung innerstädtischer amerikanischer Highways wird aufgegriffen, aber auch persifliert: „Southside, Upper City, Rest of the world." Die Betextung eines Rollfeldbildes mit Flugzeugen und Passagieren schreibt das Motto: „There's no better way to fly" in personaler Zuschreibung fort: „because of ... him, her, him, her." 2001 wirbt das Unternehmen mit „landmark"-Denkmälern europäischer Flugziele: Die Nelson-Säule in London, der napoleonische Obelisk in Paris, die Siegessäule in Berlin und ein römischer Obelisk werden „ausgepreist" mit den Flugpreisen in diese Städte. Im selben Jahr, 2001, wird das Bild eines mit der Kranich-Bildmarke geschmückten Leitwerkes, das aus den Wolken ragt, lapidar mit dem Wort „Herausragend" kombiniert. Die soziale Kompetenz und Fürsorge der Lufthansa spricht eine Werbung für alleinreisende Kinder an:

"Cool" for the other. This is sufficient to stir positive associations with Lufthansa's flights to New York. And now the sub-line reads: "You see the world the way you fly." The new tariff structures for the so-called Lufthansa Specials – special rates for business travelers – were introduced with the old graphic trick of dropping certain letters. This is a technique which mass-circulation newspapers and magazines have used for years. Photos of prominent individuals on the covers partly obscure the titles, such as *Stern*, *Spiegel* or *Brigitte*, so that they are only suggested. Recognition is guaranteed in spite of this. Now yellow becomes the major identification color for the airline. This is seen in the illustrations of a ticket stub for Marion Gräfin Dönhoff, for decades the publisher of the weekly newspaper *Die Zeit*. The comment is: "She reaches places to which we cannot take her. People's hearts." The advertisement for this "ambassador of German sensibilities in every respect," as she was once called, appeared on the occasion of her ninetieth birthday. A price-breaker advertisement for the Lufthansa Special at the turn of the millennium presented its arguments with all the typographic directness of winter closeout sales. Might this have been a return to the purely typographic designs of the mid-fifties for the new Lufthansa?

Beginning in 2000 the visual statements became more rational, although they did not completely eschew irony. Highway signposting inside American towns was shown but the text was a send-up: "Southside, Upper City, Rest of the World." The caption for a runway scene showing aircraft and passengers continues the motto – "There's no better way to fly" – in a more personalized form: "Because of … him, her, him, her." In 2001 the company advertises using landmarks in European destination cities. Nelson's Column in London, Napoleon's Obelisk in Paris, the Victory Column in Berlin and a Roman obelisk were labeled with price tags showing the air travel prices to these cities. In the same year, 2001, the photograph of a tail unit trimmed with the crane emblem and rising proudly out of the clouds is combined almost casually with the appellation "standing proud." The social competence and care typical of Lufthansa are mentioned in an advertisement for children traveling unaccompanied. Lufthansa sees more than 62,000 such youngsters safely to their destinations each year. Pure typographic advertising, announcing low, low prices, appears once again in 2002.

And so in air travel, too, just as is the case among some carmakers (the Smart here and the Maybach there), there is a gap opening up between the low-cost and the luxury sectors. It is fitting of this trend that Lufthansa, at the beginning of 2005, opened a separate lounge building at Frankfurt Airport for its newly defined and most exclusive group of passengers, the "HON Circle." It is from here that these top clients are chauffeured from airside to aircraft in Porsche SUVs or Mercedes sedans.

There are special appeals to senior citizens and the "prime of life" generation, both of which seek out and enjoy the adventures of life. These are thus specific advertisements for specific generations, advertising through a cross section of national demographics. In 2003 the technical innovations and comfort of the new Business Class seats are in the foreground while greater space is mentioned in 2004. A passenger uses binoculars to scan the newspaper held by a fellow passenger, separated from him by one empty seat. This is a photographic trick which ingeniously supports the statement, "Up to 50% more room creates 100% more personal space."

A double-page spread appearing the same year shows an amazed ten-year-old, who apparently has just discovered a fascinating surprise. In the background is a Lufthansa airplane. The young man looks at the runway. Has he discovered something wonderful just after landing or shortly before check-in? This remains unanswered and this is exactly what lends fascination to the advertisement. In any case the boy appears to

Preis-Kampagne, Ende 1990er Jahre
Price-based campaign late 1990s

Marken-Kampagne, 2004
Brand promotion campaign, 2004

1 Für die Print-Werbung der Lufthansa wurden u. a. folgende externe Agenturen für die Lufthansa tätig:
ab Ende der 1960er Jahre: DDB (Doyle, Dane, Bernbach)
1979–1985: HO & M (Heumann, Ogilvy & Mather)
1985–1989: O & M (Ogilvy & Mather)
1990–1992: SEA (Spiess, Ermisch, Abels)
1992–1996: Y & R (Young & Rubicam)
1996/97: F & L (Fanghänel & Lohmann GmbH)
1989–2000: S & J (Springer & Jacobi)
2000 bis heute: MECH (McCann-Erickson Communication House)

Preis-Kampagne, 2002
Price-based campaign, 2002

Mehr als 62.000 pro Jahr werden von ihr betreut. Auch die rein typografische Werbung, die mit Tiefpreisen wirbt, wird 2002 erneut geschaltet.

So öffnet sich gegenwärtig wie bei manchem Automobilhersteller – hier der Smart, dort der Maybach – auch im Fluggeschäft die Schere zwischen billig und exklusiv. Dazu passt, dass die Lufthansa ab Anfang 2005 für ihre neu definierte exklusivste Passagierklasse des „HON Circle" ein eigenes Lounge-Gebäude auf dem Frankfurter Flughafen in Betrieb genommen hat, von dem aus diese Top-Kunden mit Porsche-Sportwagen oder Mercedes-Limousinen zum Flugzeug chauffiert werden.

Senioren werden be- und umworben, aber auch die mittlere Generation, die die Abenteuer des Lebens genießt: spezifische Generationsansprache also, Werbung gewissermaßen als Schnitt durch die Demoskopie. 2003 stehen die Bequemlichkeit der neuen Flugsitze der Business-Klasse und ihre technische Innovation im Vordergrund, 2004 schließlich wird mit erweitertem Raumkomfort geworben. Ein Fluggast schaut mit einem Fernglas auf die Zeitung eines durch einen leer gelassenen Sitz entfernt sitzenden Mitpassagiers: eine Bildtechnik, die die Textaussage „bis zu 50 % mehr Platz schafft 100 % mehr Privatsphäre" kongenial unterstützt.

Im gleichen Jahr wird auf einer Doppelseite ein staunender Zehnjähriger gezeigt, der offenbar gerade eine faszinierende Überraschung entdeckt. Im Hintergrund erkennt man eine Lufthansa-Maschine. Schaut der Kleine auf ein Rollfeld, entdeckt er etwas Tolles nach der Landung oder kurz vor dem Einchecken? Das bleibt offen und macht gerade darum die Faszination der Werbeanzeige aus. In jedem Falle scheint der Junge ein LH-Fluggast zu sein – und das zu empfinden, was die Textzeilen betonen: „Über 120.000 Fluggäste … täglich … Das Kribbeln im Bauch, einer davon zu sein." Die visuelle Bildrhetorik entspricht in kongenialer Weise dem neu eingeführten „Claim"-Satz: „Alles für diesen Moment."

Insgesamt hat die Lufthansa-Abteilung „Werbung" weit über 800 Werbeanzeigen der letzten 50 Jahre dokumentiert, von denen hier etwa ein Zehntel in den näheren Blick genommen wurde. Nicht nur der „Werbestil" der verschiedenen beauftragten Agenturen, die insgesamt für die Lufthansa über 50 Kampagnen gestalteten, unterscheidet sich, sondern diese Kampagnen spiegeln auch den Zeitgeist, die Mode, die grafischen „Sprachen" der verschiedenen Jahrzehnte sowie die jeweiligen „Neuerungen" in der Geschichte der Lufthansa, seien sie nun technischer, flugplanstrategischer, sozialer oder ökologischer Natur, in besonders eindrücklicher Weise.

Werbung ist immer ein Spiegel der Gesellschaft, die sie hervorbringt. Sie kanalisiert und verstärkt gesellschaftliche Strömungen und hat dafür eine Indikatorfunktion. Insofern ist sie eine, wenn auch in Dienst genommene, Gattung der Kunst, die ebenso Charakter und Charisma entfalten kann wie die Malerei, der Film, die Architektur, die Fotografie oder die Literatur. Und nicht zuletzt partizipiert sie an all diesen Gattungen.

be a Lufthansa passenger – and that he feels exactly what the copy emphasizes: "More than 120,000 passengers … daily … the butterflies in the stomach that come from being one of them." The visual rhetoric in the image corresponds in an inventive way to the newly introduced motto for the campaign: "All for this one moment."

Lufthansa's Advertising Department has documented more than 800 advertisements from the last five decades, of which about one tenth have been looked at in some detail here. Not only do advertising styles differ among the various agencies who were commissioned with drafting more than fifty campaigns over the years. These advertising campaigns also reflected the *zeitgeist*, the fashions, the graphic "vernaculars" of the decades and innovations over the course of Lufthansa history – relating to technology, routing and scheduling strategies, social aspects or ecology – in a particularly impressive manner.

Advertising is always a mirror of the society from which it springs. It channels and reinforces currents within the society and is a bellwether for such currents. Thus it is a species of art, even though put into the service of commerce, which can also develop character and charisma in much the same way as painting, cinematography, architecture, photography or literature. And last but not least advertisement campaigns participate in all the mentioned genres of the arts.

**1** The advertisements for the Lufthansa were created mainly by the following agencies:
from the late 1960s: DDB (Doyle, Dane, Bern-bach)
1979–1985: HO & M (Heumann, Ogilvy & Mather)
1985–1989: O & M (Ogilvy & Mather)
1990–1992: SEA (Spiess, Ermisch, Abels)
1992–1996: Y & R (Young & Rubicam)
1996/97: F & L (Fanghänel & Lohmann GmbH)
1989–2000: S & J (Springer & Jacobi)
2000 until today: MECH (McCann-Erickson Communication House)

Neue Business-Class-Kampagne, 2004
New Business Class campaign 2004

Kampagne für Fernreisen, 1963
Campaign featuring distant destinations, 1963

Kampagne mit Mitarbeitern, 1975
Campaign featuring employees, 1975

Kompetenz-Kampagne, 1979–1981
Competence campaign, 1979–1981

Marken-Kampagne, 1983–1985
Brand promotion campaign, 1983–1985

„Wolken-Kampagne", ab 1994
"Clouds campaign," as of 1994

„Wolken-Kampagne", ab 1994
"Clouds campaign," as of 1994

Kampagne zur Einführung des neuen Kabinen-Interieurs, 1998
Campaign announcing the introduction of new cabin interiors, 1998

Image-Kampagne, 1998
Image campaign, 1998

Image-Kampagne, 2001
Image campaign, 2001

Preis-Kampagne, 2002/2003
Price-based campaign, 2002/03

„Vertrauenskampagne", 2004
"Customer confidence campaign," 2004

Jubiläumsanzeige 50 Jahre Lufthansa, 2005
Anniversary advertisement – Lufthansa turns 50, 2005

LUFTHANSA

# „Weißt Du noch, damals ...“

Der Begriff „Erinnerungskultur“ meint jene Dinge, die mit persönlichen Erlebnissen und Ereignissen verbunden sind. Dazu gehören z. B. individuelle Dankschreiben und Urkunden für langjährige Betriebszugehörigkeit, Ehrennadeln, aber auch Rangabzeichen, Tressen und Litzen auf den Uniformen der Lufthansa, Ersttagsbriefe für neu aufgenommene Flugverbindungen sowie Werbegeschenke, Aufkleber und Revers-Nadeln. Ebenso aber gehören im prinzipiellen Unternehmenszusammenhang grafisch gestaltete Erzeugnisse wie etwa Speisekarten, Plakate oder Kalenderblätter dazu, sobald sie, z. B. durch eine persönliche Widmung auf einen individuellen Empfänger hin personalisiert werden.

Dass Gegenstände zu Objekten einer Erinnerungskultur werden können, hängt allerdings weniger von der Art und Weise ihres Erwerbs ab als von den affektiven Bindungen, die sie auslösen. Urkunden kann man überreicht bekommen, aber auch auf Auktionen erwerben; Ersttagsbriefe selber versenden oder gesendet bekommen; Anstecknadeln als Zeichen der realen Zugehörigkeit zu einem Unternehmen tragen oder im Nachlass finden; Werbegeschenke in Flugzeugen oder Flughäfen als Geschenk erhalten oder aber sammeln. Auch wenn eine Speisekarte, etwa mit einer persönlichen Widmung eines prominenten Mitpassagiers, ursprünglich eine spezielle Person meint, kann ein solches Objekt auch für andere erinnerungskulturell bedeutsam werden. Die Affektfunktionen solcher Objekte realisieren sich in verschiedenen Bedeutungskontexten: das LH-Give-away „Bommmerlunder-Steingutfläschchen“ mag an einen Flug oder an einen Flohmarktbummel erinnern, bei dem man es erworben hat.

„Erinnerungskultur“ beschreibt aber auch eine zeitliche Distanz. Erinnern kann man sich nur an zurückliegende Ereignisse und diese werden dann mit Hilfe von Gegenständen in der Gegenwart reaktiviert: als „Memento“ an Ereignisse, Orte, Situationen, Personen. Dabei bekommen die Dinge Symbolcharakter, sie gemahnen in jeweils aktueller Gegenwart an Vergangenes. Und diese Objekte der Erinnerungskultur erlangen nicht selten Trophäen- bzw. Souvenircharakter, werden gesammelt, gekauft, getauscht: Börsen, Auktionen, Versteigerungen sind die Distributionsorte für diese Sammelbedürfnisse. Zur „Erinnerungskultur“ gehören aber auch verbale Erzählungen darüber, „wie es einmal war“: Die Geschichtswissenschaft nennt dies „oral history“. Es gehört zu einem runden Bild der Darstellung der „Gegenstandswelten“ eines Unternehmens, auch solche persönlichen Aspekte zu berücksichtigen, weil auch sie ebenso symptomatisch wie charakteristisch für eine Firmenkultur sind.

So sind etwa im Kölner Archiv der Lufthansa Dutzende von „Erstflug-Briefumschlägen“ dokumentiert, die diverse Premieren neuer Flugverbindungen per Sonderstempel und manchmal durch Briefmarken mit dem Kranich-Logo der Fluggesellschaft den jeweiligen Empfängern mitteilen. Sie sind auf den Tag genau datierbar, rare Philatelisten-„Schnäppchen“. Ersttagsstempel dokumentieren den Stempelort, den Erscheinungstag sowie den Anlass der Marke und des Ereignisses. Seit 1955 wurde jeder Erstflug der Lufthansa philatelistisch begleitet.

Es beginnt mit einem Luftpostumschlag, der den ersten Flug der Deutschen Lufthansa, also die Wiederaufnahme des zivilen Luftverkehrs nach dem Krieg am 1.4.1955, dokumentiert. Auch das Streckennetz mit den ersten fünf angeflogenen Flughäfen Hamburg, Düsseldorf, Köln/Bonn, Frankfurt am Main und München ist auf dem Couvert abgebildet. Die vier Briefmarken zeigen das Kranich-Logo mit den Nennwerten fünf, zehn, 15 und 20 Pfennig, gedruckt im Offsetdruck und mit Wasserzeichen versehen, mit

Seite 208: Aschenbecher, um 1960
Page 208: Ashtray, ca 1960

# "Do you remember when ...?"

The concept behind the "culture of recollection" covers all those things associated with personal experiences and events. Examples of this might be thank-you notes and certificates attesting to many years of company service, service pins, insignia of rank, braid or cord and stripes on the Lufthansa uniforms, "first day covers" for newly instituted routes, and advertising gifts, stickers and lapel pins. Among other such items within the fundamental corporate context are products with a graphic design – such as menus, posters or calendar sheets whenever they are, for instance, personalized for the individual recipient with a personal note or dedication.

Whether physical objects become part of a culture of recollection depends less on the way in which they were acquired than on the emotional ties associated with them. Certificates may be presented to the recipient but they can also be acquired at auctions; first day covers can be sent or received; one might wear a service pin as a sign of association with a company or find one in the estate left by deceased parents; advertising gifts could be received in the airplane or at the airports or be collected at flea markets. Even where a menu, perhaps autographed by a prominent fellow passenger, was initially intended for a specific person, such an object can also have significance as memorabilia for others. The emotional functions of objects such as these are realized in various contexts. The "Bommerlunder stoneware bottle," a Lufthansa giveaway, might remind its owner of a flight or strolling through the flea market where it was purchased.

A "culture of recollection" also requires a certain passage of time. One can only remember past events and these are then reactivated in the present with the help of physical objects – mementos of events, places, situations or people. Here these items acquire a symbolic character; operating in the present, they remind us of things past. And it is not seldom that these objects of recollection attain the status of trophies or souvenirs – they are collected, bought and sold, traded. Collectors' exchanges and auctions are the marketplaces for these objects of collectors' desire. Also a part of a culture of recognition is recounting verbally, "how things used to be." The social sciences refer to this as oral history. It complements and rounds out the depiction of any company's "world of physical objects." Personal aspects such as this need to be considered because they are both symptomatic for and characteristic of any corporate culture.

Thus dozens of "first day covers" are filed away at the Lufthansa archives in Cologne. They tell their recipients about the debut flights on new routes – with a special cancellation stamp and sometimes postage stamps bearing the airline's crane symbol. They can be dated to the exact day and thus are rare philatelic treasures. First-day cancellations document the cancellation location, the day of issue, the reason for the mark and the event. Ever since 1955 every first flight inaugurated by Lufthansa has been commemorated in this way.

The collection starts with an air mail envelope which documents the very first flight by the Deutsche Lufthansa, marking the resumption – following the end of the World War II – of civil air traffic on April 1, 1955. The route network indicating the first five airports to be served – Hamburg, Düsseldorf, Cologne/Bonn, Frankfurt and Munich – is also reproduced on this cover. The four postage stamps show the crane logo and are in denominations of five, ten, fifteen and twenty pfennigs. They are offset printed and incorporate a watermark; a shadow effect lends dynamic character. Also affixed to the envelope is a blue tax stamp bearing the inscription "Emergency relief for Berlin." Its

Ersttagsbrief: Aufnahme des Europaverkehrs, 1955
First-day cover: Launch of European service, 1955

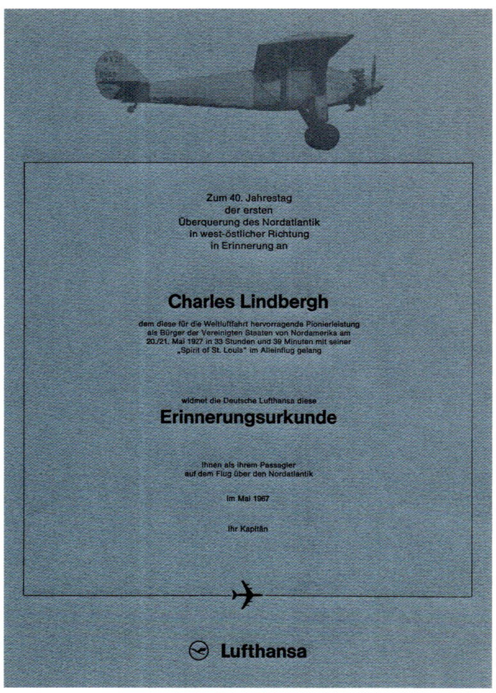
Zum 40. Jahrestag
der ersten
Überquerung des Nordatlantik
in west-östlicher Richtung
in Erinnerung an

**Charles Lindbergh**

dem diese für die Weltluftfahrt hervorragende Pionierleistung
als Bürger der Vereinigten Staaten von Nordamerika am
20./21. Mai 1927 in 33 Stunden und 39 Minuten mit seiner
„Spirit of St. Louis" im Alleinflug gelang

widmet die Deutsche Lufthansa diese

**Erinnerungsurkunde**

Ihnen als Ihrem Passagier
auf dem Flug über den Nordatlantik

im Mai 1967

Ihr Kapitän

Lufthansa

Erinnerungsurkunde zum 40. Jahrestag der Atlantiküberquerung, 1967
Certificate commemorating the 40th anniversary of the first non-stop Atlantic crossing, 1967

dynamisch hinterlegtem Schatten. Zusätzlich klebt auf dem Umschlag die blaue Steuermarke „Notopfer Berlin" mit dem Nennwert von zwei Pfennig. Mit den gleichen vier Postwertzeichen sind auch zwei weitere Ersttagsbriefe zur Aufnahme des Europaverkehrs mit Verbindungen nach London, Paris und Madrid am 15.5.1955 frankiert. Am 1.4.1960 folgt der tägliche Non-Stop-Flug von Frankfurt am Main nach New York, vorgestellt mit einem in Preußischblau und Sonnengelb gedruckten Signet, das einen Boeing-Jet über der Skyline Manhattans zeigt. Sechs Wochen später folgt der Erstflug von San Francisco via Montreal und Paris nach Frankfurt am Main: Erst gut acht Jahre später folgt der Erstflug in umgekehrter Richtung. Ein gutes Jahr später, 1961, findet der erste Lufthansa-Flug mit einer Boeing 707 von New York nach Köln und München statt. Den entsprechenden Umschlag ziert das Bild einer Passagierin, die von einem smokingbewehrten Steward gerade ein luxuriöses Buffet serviert bekommt. Mitte 1962 folgt die Südafrikastrecke mit dem Flug nach Khartum. Noch wird die Wortmarke „Lufthansa" in Großbuchstaben gedruckt. Der Erstflugbrief zur Direktverbindung New York–Stuttgart vom 25.4.1964 zeigt dann bereits die Wortmarke in Groß- und Kleinschreibung mit dem Kranich im Kreis; ebenso ein Umschlag, der den ersten Flug über den Nordpol nach Tokio via Hamburg, Kopenhagen und Anchorage, der am 28./29.5.1964 stattfand, mit dem Signet eines klassisch-japanischen Tempelschreins annonciert.

Neben Ersttagsbriefen gab es zu vergleichbaren Anlässen ab und zu auch Erinnerungsplaketten, z. B. 1962 zur Eröffnung des West- und Südafrikadienstes nach Lagos bzw. Johannesburg, aber auch Glückwunschschreiben von Mandatsträgern innerhalb und außerhalb der Lufthansa.

In den Bereich der Urkunden fällt eine Reproduktionsvorlage für ein- bzw. zweifarbige Zeichen, die auf Dokumenten für zehnjährige Betriebszugehörigkeit Verwendung finden sollten. Auffällig ist, dass diese Signets aus Text und Zahlensymbol, 1965 von Otl Aicher entworfen, in ihrem Spiel mit Positiv- und Negativflächen deutlich Anklänge an die Pop-Art-Zahlen und Buchstabenbilder des amerikanischen Pop-Art-Malers Robert Indiana aus der gleichen Zeit zeigen. Konservativer wirkt dann die tatsächliche Urkunde für „Zehn Jahre treue Mitarbeit". Im Mai 1967 widmete „Ihr Kapitän" den Passagieren auf

nominal value is two pfennigs. It was with the same four postage stamps that two further first-day covers were franked; they marked the May 15, 1955, commencement of intra-European service, with routes to London, Paris and Madrid. This was followed some years later, on April 1, 1960, by a daily, non-stop flight from Frankfurt to New York, presented with a signet printed in Prussian blue and sunshine yellow and showing a Boeing jet over the Manhattan skyline. This was followed just six weeks later by the first flight from San Francisco, via Montreal and Paris, to Frankfurt. The first flight in the opposite direction was, however, not inaugurated until a good eight years later. In 1961 a Boeing 707 lifted off from New York City, destined for Cologne and Munich. The envelope printed for this occasion shows a steward in formal attire serving a luxurious buffet to a passenger. Service to southern Africa followed in mid-1962 with the flight to Khartoum. Here the Lufthansa logotype is still printed all in capitals. The first day cover commemorating the non-stop link from New York to Stuttgart on April 25, 1964, then shows the logotype in standard capitalization, together with the crane inside the circle, which is also the case for a cover prepared for the first flight over the North Pole, to Tokyo, via Hamburg, Copenhagen and Anchorage. This route, first flown on April 28/29, 1964, was announced with the image of a classical Japanese shrine.

In addition to first-day covers, commemorative posters are occasionally printed for noteworthy occasions such as the institution of service to western and southern Africa in 1962, with routes to Lagos and Johannesburg. Also to be included in this category are congratulatory letters sent to dignitaries inside and outside the Lufthansa organization.

To be found in the field of certificates and testimonials is the artwork for a single- or two-color symbol which was to have been used on documents certifying ten years of service to the company. Quite noticeable is that this signet, comprising both copy and numbers, was drafted by Otl Aicher in 1965. Its experimentation with positive and negative areas shows the major influence exerted by works of art incorporating numbers and letters and created by his contemporary, American pop art painter Robert Indiana. The actual certificate for "Ten years of loyal service" turned out to be a bit more conservative. In May of 1967 "your captain" on the flight across the North Atlantic presented to passengers a "Commemorative Certificate" recalling the fortieth anniversary of the first

Urkunde
Die Deutsche
Lufthansa AG
und
Lufthansa Service GmbH
danken

für 10 Jahre
treuer Mitarbeit

Die Geschäftsleitung

Urkunde für zehnjährige Firmenzugehörigkeit, 1965
Certificate honoring 10 years of service, 1965

Plakette Eröffnung des West- und Süd-Afrika-Dienstes, 1962
Medallion marking the commencement of service to West and Southern Africa, 1962

Klappmesser, Ende 1950er Jahre
Pocket knife, late 1950s

dem Flug über dem Nordatlantik eine „Erinnerungsurkunde" zum 40. Jahrestag des ersten Nordatlantikfluges, der Charles Lindbergh mit seiner „Spirit of St. Louis" im Mai 1927 in über 33 Stunden gelang.

Zur Erinnerungskultur gehören aber auch Werbegeschenke wie etwa ein besonders schönes Klappmesser aus lackiertem Metall, bei dem das Leitwerk haiflossenähnlich als Kapselheber ausgebildet ist. Anmutung und Form verweisen dieses Klappmesser in die Anfangsjahre der Lufthansa. Auch eine in Leder gebundene Schatulle in Buchform mit einer eingeprägten Lockheed Super Constellation und dem Versalienschriftzug der Fluggesellschaft sowie den noch kreislosen Kranich muss deshalb aus der Zeit vor 1962 stammen. Im gleichen Zeitraum werden dem Fluggästen der Senator-Klasse einzeln in Metallröhrchen verpackte Zigarren angeboten. Auf ein paar Jahre später, etwa auf die Mitte der 1960er Jahre, lässt sich das Werbegeschenk eines „Brandenburger Tores" als Zinnmodell in einer aufklappbaren Geschenkkassette mit deutschem und englischem Erläuterungstext datieren, ebenso wie ein Kugelschreiber oder ein ledernes Visitenkartenetui, beides mit der LH-Wortbildmarke. Warum allerdings zum Gedenken an die Eröffnung des „Lufthansa Service zwischen Houston, Texas, und Frankfurt am Main" am 2.7.1985 ausgerechnet ein irdener Maßkrug als Werbegeschenk dienlich schien, mag man der Hofbräuhaus-Nostalgie amerikanischer Touristen zuschreiben. Ähnlich nostalgisch wirkt das Bommerlunder-Schnapsfläschen in glasierter Keramik im norddeutschen Bauernlook, das um 1970 als Werbegeschenk genutzt wurde; mit durchaus unterschiedlichen Motiven. Senftöpfchen, Salz- und Pfefferstreuer kamen hinzu. Ein dazu passendes Tischtuch zeigt altdeutsche Spielkartenfiguren und als Rahmung eine serielle Reihung der Wortbildmarke Lufthansa. Auch Schallplatten und Musikkassetten dienten der Kundenbindung, von frühen Langspielplatten der 1950er Jahre mit Musik etwa des Tanzorchesters Max Greger über Schallplatten für Kinder, z. B. „Die Biene und das Düsenflugzeug", bis zu Hörkassetten mit „Classic Masterpieces" oder auch schon mal dem Thema „Fitness in the air". Früh entdeckte die Lufthansa zudem den Nachwuchs als Kundschaft. Sie entwickelte einen sogenannten „Junior Service" und überreichte den juvenialen Frühfliegern eine Mappe mit Logbuch, Anstekker und Ausweis. Der Klappentext, natürlich im vertraulichen „Du" gehalten, verdeutlicht die Intention: „Lieber Junior, wir freuen uns, Dich im Kreis unserer Lufthansa-Junioren zu begrüßen. Hiermit überreichen wir Dir Deinen Ausweis und Dein Abzeichen, an dem man Dich gleich als Lufthansa-Junior erkennt. Dein persönliches Logbuch, in dem Du Einzelheiten über Deine Reisen mit Lufthansa festhalten kannst. Wir hoffen, dass Dir Deine nächsten Lufthansa-Flüge mit dieser Ausrüstung noch mehr Spaß machen. Auf baldiges Wiedersehen an Bord. Lufthansa."

Das hat schon eine gewisse Chutzpe. Ebenso gehörten zum „Junior Service" ein Memoryspiel mit historischen und aktuellen Flugzeugmotiven sowie Würfelspiele, weitere Spielkarten und Geschicklichkeitsspiele.

Polydor: Schallplattte Lufthansa-Konzert, 1960er Jahre
Polydor: Lufthansa Concert LP, 1960s

Auch Werbeaufkleber erfreuten sich von Anfang an ebenso wie vor dem Krieg erneut großer Beliebtheit. Um 1955 zeigen sie grafisch vereinfachte Globussymbole, verknüpft und verschränkt mit dem Begriff der Lufthansa, was offenbar so viel heißen will wie: Wir sind wieder wer! Auch im fernen Ausland wird dieser grafisch-nüchterne Stil beibehalten, aber schon auch einmal folklorisiert, wie im fernen Südamerika. In den späten 1960er Jahren produzierte die Fluglinie entsprechend dem Zeitgeist flippig-ironische Aufkleber, die der damals grassierenden Button-Mentalität entsprachen. Wortspiele, Gesinnungsmetaphern gewissermaßen waren en vogue. Da hieß es etwa: „Ich flieg gern weg mit einem Kranich am Heck", „Mein liebster Fimmel mit Lufthansa am Himmel", „Bei Lufthansa hebt man ab" oder einfach „Lufthansa. The better choice". Der

non-stop solo flight across the North Atlantic, which Charles Lindbergh accomplished with his Spirit of St. Louis in over 33 hours in May of 1927.

Also a part of the culture of recollection are advertising gifts such as a particularly attractive pocketknife made of enameled metal. where the tail unit takes the shape of a shark fin and functions as a bottle opener. The shape and feel assign this pocketknife to Lufthansa's early post-war years. A leather-bound jewelry case in the form of a book, embossed with the image of the Lockheed Super Constellation and decorated with the all-caps airline logotype together with the crane *sans* circle, must date back to pre-1962. During the same timeframe passengers in the Senator Class were offered cigars individually packed in metal tubes. A tin model of the Brandenburg Gate in a hinged gift box, with explanatory text in German and English, can be dated to a few years later, about the middle of the 1960s. This is also true for a ballpoint pen and a leather calling card case, both with the Lufthansa emblem and logotype. The selection of an earthenware tankard as an advertising gift to commemorate the opening of Lufthansa service between Houston, Texas, and Frankfurt on July 2, 1985, might be attributed to the Hofbräuhaus nostalgia so often embraced by American tourists. Similarly nostalgic is the effect of the Bommerlunder schnapps bottle in glazed ceramic, reflecting the northern German country look, used around 1970 as an advertising gift with a variety of motifs. Mustard jars and salt and pepper shakers joined them. A matching table cloth shows old German playing card figures, edged with a series of Lufthansa emblems and logotypes. Phonograph records were also employed to promote customer loyalty, beginning with the early LPs of the 1950s featuring music played by the Max Greger dance band, or children's records such as "The Bee and the Jet Plane." These were later followed by audio cassettes with "Classic Masterpieces" and even occasionally covering "Fitness in the Air." Lufthansa soon discovered young people as customers. The company developed the so-called "Junior Service" and presented to the youthful fliers a folder containing a logbook, a pen and a traveler's ID card. The copy on the inside of the front cover makes the intention clear.

Zigarrenkiste Senator-Klasse, Anfang 1960er Jahre
Senator Class cigar box, early 1960s

"Dear Junior, We are happy to welcome you to the 'Lufthansa Juniors.' We are presenting you your ID card and pin which will immediately identify you to others as a Lufthansa Junior. You can record the details of your Lufthansa flight in your personal log book. We hope that these accessories will make your next Lufthansa flight even more fun. We look forward to welcoming you on board again soon. Lufthansa."

That shows a fair degree of chutzpa. Also part of the Junior Service is a memory game with historical and current aircraft motifs, dice games, additional playing cards and games of skill.

Advertising stickers also enjoyed great popularity, just as they had before the war. Around 1955 they showed graphically simplified symbols of the globe, linked with and crossed by the word Lufthansa, apparently intending to say: "We're back again!" This sober graphic style is also retained in distant countries even though a touch of local color was sometimes added, as was seen in South America. In the late 1960s the carrier produced flippant, ironic stickers which reflected the button craze so prevalent at that time. Plays on words and ideological metaphors were in vogue. There we see, for example: "In the clouds I love to sail, with a crane upon the tail," "Lufthansa in the skies – crazy and nice," "Fly away with Lufthansa" or simply "Lufthansa. The better choice." A sticker which almost casually places the question "What else?" beneath a photo of a Lufthansa tail unit with the crane in the circle also reflects Germany's pop culture of the time, which often used English expressions. The 1970s were characterized by something of a punctuation mania, setting a mark after every term. This is seen on a

Lufthansa Junior Service: Logbuch, Anstecker, Ausweis, Mitte 1960er Jahre
Lufthansa Junior Service: Logbook, pin, ID card, mid-1960s

mit Anglizismen operierenden Popkultur der Zeit entspricht auch jener Aufkleber, der unter das Foto eines LH-Leitwerkes mit Kranich im Kreis lapidar die Frage setzt: „What else?“ Den 1970er Jahren mit ihrer Satzpunkt-Euphorie nach jedem Begriff entspricht ein Werbeaufkleber, der in fett gedruckter Otl-Aicher-Helvetica auf fast durchgängig preußischblauem Rechteck mit einer klein abgebildeten, im Flug befindlichen Maschine markig statuiert: „Meine Linie. Lufthansa.“ Aber offenbar eignen sich Aufkleber auch für die persönliche Ansprache. 1998 werden die Mitarbeiter der „Lufthansa CityLine“ zum Jahreswechsel mit einem Aufkleber bedacht, auf dem zu lesen steht: „Herzlichen Dank für Ihren Einsatz und Ihre hohe Leistungsbereitschaft im Jahr 1998. Einen ‚spritzigen' Start ins neue Jahr wünscht Ihnen Ihre Geschäftsführung und das Leiter-Team.“

Kranich-Anstecker, Mitte 1960er Jahre
Crane pin, mid-1960s

Auch wenn sie nicht so personalisiert sind wie Urkunden oder Erstflugbriefe, sind die Anstecker für die Dienstbekleidung der Lufthanseaten ebenso wie die Rangabzeichen für das fliegende Personal doch auch Symbole persönlicher Zugehörigkeit zum Unternehmen. In den ersten Jahren gibt es vermessingte Kranich-Anstecker noch ohne den Kreis, als reines Emblem also, oder mit integriertem Namensschild. Es folgen der Kranich im Kreis als offene Konstruktion und dann verschiedene emaillierte Anstecker in Form geschlossener Kreise, die das Logo entweder in Preußisch- oder Himmelblau auf Violett, Türkis oder Chamois zeigen. Naturalistische Flugzeugmodelle fürs Revers ergänzen diese Ansteckerpalette, aber auch jeweils doppelte Leitwerkromben, die jeweils das LH-Emblem mit einem weiteren Logo der „Star Alliance“-Partner kombinieren. Hinzu kommen rechteckige Anstecker mit Lufthansa-Logo und dem generellen, von Pentagram entwickelten „Star Alliance“-Logo sowie comichafte Jumbo Pins.

Schreibtischuhr, 2002/03
Desk clock, 2002/03

Und auch jene Poster, die der WorldShop der LH anbietet – Flugzeuge in der Luft, Cockpit-Aufnahmen, Weltkarten –, sind so wie die ebenfalls erwerbbaren Nachdrucke früher Werbeplakate ein Indiz für eine Strategie, dem Unternehmen werbewirksam Historie und Tradition zu bescheinigen: eine Bescheinigung allerdings, der dieses Unternehmen wohl kaum bedarf. Aber darum geht es auch nicht: Der Kauf eines Nostalgieposters der LH ist insofern auch deutlich mehr einer „Ironie des Einverständnisses“ als einer „Neugier historischer Korrektheit“ geschuldet. Jene Schreibtischuhren allerdings, die runde Präzisionszifferblätter in „original poliertem Flugzeugaluminium“ einsetzen, partizipieren deutlich am Gout der „Außerdienststellung“. Einem solchen Schreibtischutensil haftet die Aura der Stratosphäre an: kalt und lebensfeindlich. Zurückgeholt auf die Erde, domestiziert, personalisiert: Auch da wird High-Tech zu High-Touch!

Erinnerungen sind vielfältig: persönlicher, sozialer, psychologischer, technischer, kommunikativer Struktur. Die Psyche braucht die Krücke, die Brücke physischer Objekte: Nur so bekommen Erinnerungen ihren Fokus. Und so verschränkt die „Erinnerungskultur“ aller Unternehmen und auch jene der Lufthansa private Memorabilien mit öffentlich-institutionellen Elementen, die auch zur „Corporate Identity“ insgesamt beitragen. Aber die Rolle dieser unterschiedlichen Elemente im emotionalen Haushalt der Menschen bleibt gleich: Es sind Indikatoren für die personalisierte, individualistische Wahrnehmung eines Unternehmens; eine „In-house“- und Mitarbeiterperspektive, die auch über den Erwerb solcher Gegenstände Kunden psychologisch zu Firmenangehörigen macht. „Erinnerungskultur“ meint insofern: Wir sind eine Familie!

rectangular advertising sticker, Prussian blue almost throughout, showing an airplane in flight. The boldfaced Otl Aicher Helvetica copy declares: “My line. Lufthansa.” But stickers are apparently also suitable for more personalized messages. Toward the end of 1998 Lufthansa CityLine employees received as a small memento a sticker reading: “Heartfelt thanks for your dedication and your willingness to serve in 1998. Your executives and the management team want to wish you a ‘sparkling’ start for the New Year.”

Even though they are not as personal as certificates or first-flight covers, the pins for the Lufthansa service uniforms and the insignia of rank for the flight and cabin crews are also symbols of an individual’s affiliation with the company. In the early years there were brass-plated crane pins, still without the circle, either as a stand-alone emblem or integrated into a nameplate. This was followed by the openwork crane in the circle and then various enameled pins in the form of solid circles which bore the emblem either in Prussian or sky blue on violet, turquoise or chamois. Realistic aircraft miniatures to be worn on the lapel augmented this line of pins, as did tandem tail units which combined the Lufthansa emblem with a further logo from one of the other Star Alliance partners. Also found were rectangular pins with the Lufthansa logo and the general Star Alliance logo developed by Pentagram along with comic-like “Jumbo Jet” pins.

Those posters which the Lufthansa WorldShop offers – aircraft in the air, cockpit photos, world maps – just like the reprints of earlier advertising posters are also an indication of a strategy which confirms the history and tradition of the company with good advertising effect. This is a kind of authentication which the company hardly needs. But that’s not the issue at hand here. Purchasing a nostalgic Lufthansa poster owes significantly more to an “irony of acquiescence” than to “curiosity about historical correctness.” The desk clocks which use round, precision dials in “original polished aircraft aluminum” clearly express a sense of “decommissioning.” A desk utensil like this has something stratospheric about it: cold and inhospitable. When it is pulled back to earth, domesticated and personalized, then high-tech once again becomes high-touch!

Memories are many-faceted. Their structures may be personal, social, psychological, technical or communicative. The psyche needs a crutch – a bridge formed by physical objects. Only in this way are memories given focus. And so the “culture of recollection” in every company, and at Lufthansa, too, intertwines private memorabilia with public and institutional elements which also contribute overall to corporate identity. The role played by these differing elements in the human emotional balance nonetheless remains the same. These are indicators for a personalized, individualized perception of a company. This is an “in-house” and employee perspective which makes customers, through the acquisition of such objects, a part of the corporate family, at least psychologically. The “culture of recollection” thus means: “We are all members of one family!”

Anstecker mit LH-Logo, 1960er Jahre
Pin with Lufthansa logo, 1960s

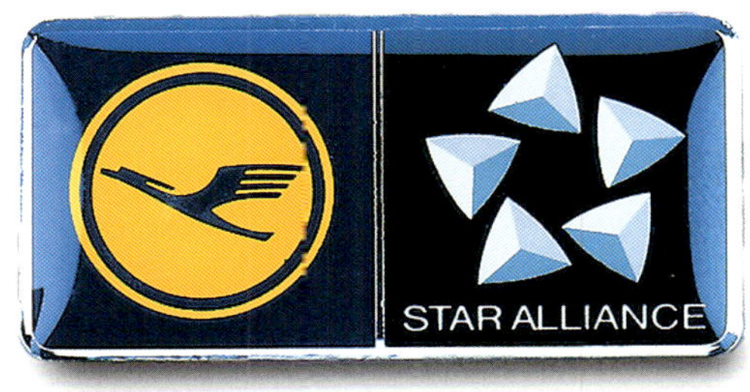


Pin LH/Star Alliance, Ende 1990er Jahre
Lufthansa/Star Alliance pin, late 1990s

Bommerlunderfläschchen, ca. 1970
Bommerlunder bottle, about 1970

Bierkrug zur Aufnahme der Direktverbindung nach Houston/Texas, 1985
Tankard marking commencement of direct service to Houston, Texas, 1985

Schnapsfläschchen, Mitte 1960er Jahre
Schnapps flask, mid-1960s

Schnupftabakfläschchen, 1960er Jahre
Snuff bottles, 1960s

Schnapsfläschchen, Mitte 1960er Jahre
Schnapps flask, mid-1960s

Schnupftabakfläschchen, 1960er Jahre
Snuff bottle, 1960s

Schnapsfläschchen, Mitte 1960er Jahre
Schnapps flask, mid-1960s

Aufkleber, ca. 1955
Sticker, about 1955

Aufkleber Südamerika, Mitte 1960er Jahre
Sticker, South America, mid-1960s

Aufkleber, ca. 1955
Sticker, about 1955

Aufkleber, ca. 1955
Sticker, about 1955

Aufkleber „Meine Linie", Anfang 1980er Jahre
"My airline" sticker, early 1980s

Aufkleber, 1970er Jahre
Sticker, 1970s

Ersttagsbrief : Aufnahme des Flugverkehrs San Francisco–Frankfurt, 1960
First-day cover: Start of service San Francisco–Frankfurt, 1960

Ersttagsbrief: Aufnahme des Liniendienstes Hamburg–Paris, 1965
First-day cover: Beginning of scheduled service Hamburg–Paris, 1965

Ersttagsbrief: Aufnahme des Flugverkehrs Frankfurt–Khartoum, 1976
First-day cover: Start of service Frankfurt–Khartoum, 1976

Ersttagsbrief: Aufnahme des Flugverkehrs Frankfurt–Mauritius, 1970
First-day cover: Start of service Frankfurt–Mauritius, 1970

Ersttagsbrief: Aufnahme des Flugverkehrs Frankfurt–Melbourne, 1980
First-day cover: Start of service Frankfurt–Melbourne, 1980

LUFTHANSA
D-AIRX

**Ausgewählte Literatur**

**Aicher, Otl:** „flugapparate von paul mac cready“ (1981), Wiederabdruck in: *form*, Nr. 199, Nov./Dez. 2004, S. 53 ff.
**Anonym:** „Das gelb-weiße Designkonzept ist durchgeflogen“, *Design Report*, Nr. 9, 1989, S. 13 ff.
**Anonym:** „Freudvolle Formen als Ausdruck sachlicher Funktionalität. Wolf Karnagel gestaltete für die Deutsche Lufthansa die neue Bordausstattung“, *form*, Nr. 116, 1986, S. 25 ff.
**Anonym:** „Lufthansa – die konsistente Entwicklung einer Corporate Identity“, in: Birkigt/Stadler/Funck (Hrsg.) *Corporate Identity – Grundlagen, Funktionen, Fallbeispiele*, Frankfurt am Main 2002, S. 421 ff.
**Bachinger, Richard:** *Unternehmenskultur – Ein Weg zum Markterfolg*, Frankfurt am Main 1994.
**Coutts Clay, Jennifer:** „Ways ahead“, in: *Jetliner Cabins*, London 2003, S. 177 ff.
**Deutsche Lufthansa AG (Hrsg.):** *Follow me*, Köln, o. J (1980er Jahre).
**Deutsche Lufthansa AG (Hrsg.):** *Wir fliegen wieder nach Berlin*, Köln 1990.
**Deutsche Lufthansa AG (Hrsg.):** *Die Zeit im Fluge – Geschichte der Lufthansa seit 1926*, Frankfurt am Main 2002.
**Edelmann, Klaus Thomas:** „Irrflug?“, *Design Report*, Nr. 9, 1989, S. 12.
**Eisenbrand, Jochen; Vegesack, Alexander von (Hrsg.):** *Airworld – Design und Architektur für die Flugreise*, Ausstellungskatalog, Vitra Design Museum, Weil am Rhein 2004.
**Esslinger, Hartmut:** „Kranich auf neuem Kurs“, *form spezial*, Nr. 2, 1998, S. 58 ff.
**Franck, Bernd:** „Und wie man sich zeigt, so ist man. Rückblick. Vor 35 Jahren verpasste Otl Aicher der Lufthansa ein neues Erscheinungsbild“, *form spezial*, Nr. 2, 1998, S. 68 f.
**Geberzahn, W. O., und Grützner, Juliane:** „Mode, Mysterium und Design. Gabriele Strehle formt das Label Strenesse zu einer internationalen Marke“, *Design Report*, Nr. 10, 2001, S. 52 ff.
**Harms, Ingeborg:** „Farben des Himmels. Pop-Kraniche, Orient-Formel und Hedonisten-Schick: Die Mode der Flugbegleiter. Eine Stilkritik“, *Frankfurter Allgemeine Sonntagszeitung*, 30. 6. 2002, S. V2, V3.
**IF International Forum Design GmbH (Hrsg.):** *Designpreis Industrie Forum Hannover 2002*, Hannover 2002.
**Jens, Tilman:** „Des Kranichs neue Kleider“, *Design Report*, Nr. 12, 1994, S. 34 f.
**Jens, Tillman:** „Unter den Wolken: Ein notorischer Vielflieger erlebt das neue Lufthansa Design“ *form spezial 2*, 1998, S. 66 f.
**Knies, Daniel:** „Flugzeuge im Bauch – Essen über den Wolken“, in: Hochschule für Gestaltung Offenbach am Main (Hrsg.): *auch anders – Jahresbericht 2004*, Offenbach am Main, 2004, S. 226.
**Loschek, Ingrid:** *Reclams Mode- und Kostümlexikon*, Stuttgart 1999.
**Lovegrove, Keith:** *Airline. Identity, Design and Culture*, New York 2000, S. 109, 126 f.
**Mies, Petra:** „Mode zum Abheben“, *Frankfurter Rundschau Magazin*, 5. 1. 2002.
**Müller, Birgit:** „Abenteuerlicher Auftritt“, *Design Report*, Nr. 5, 1997, S. 39.
**Prinz, Susanne:** *Besteck des 20. Jahrhunderts. Vom Tafelsilber zum Wegwerfartikel*, München 1993, S. 71 f.
**Rahskopff, Hermann:** Gedächtnis-Protokoll nach einem Besuch bei Otto Firle in Düsseldorf am 18. 11. 1965, LH-Archiv, Frankfurt am Main.
**Schmidt, Doris:** „Geschirr in der Luft“, in: *Zeitgemäße Form. Industrial Design international*, München 1967, S. 91.
**Schweer, Michael:** „Die richtige Kurve ist seine Stärke“, *md*, Nr. 2, 1991, S. 49.
**Schweer, Michael:** „Hightech in Form. Müller/Romca Industriedesign im Porträt“, *md*, Nr. 7, 2004, S. 52 ff.
**Steguweit, Christian:** „Corporate Identity – Problem oder Lösung für die Fragen der 90er Jahre – Eine Standortbestimmung am Beispiel der Lufthansa“, in: Norbert Hammer: *Die stillen Designer – Manager des Designs*, Essen 1994, S. 103–116.
**Steguweit, Christian:** „Lufthansa. Typologie und Konsequenz einer Corporate Identity“, in: Klaus Schmidt (Hrsg.): *Corporate Identity in Europa: Strategien, Instrumente, Erfolgreiche Beispiele*, Frankfurt am Main / New York 1994, S. 135–142.
**Sweet, Fay:** *frog – form follows emotion*, New York 1999.
**Willich, Dieter:** „Vom Ursprünglichen und Ähnlichen … oder die Quadratur des Plagiarius: Baumann gegen Karnagel, Karnagel gegen Baumann, Baumann gegen Willich“, *form*, Nr. 130, 1991, S. 64–67.

**Selected bibliography**

**Aicher, Otl:** “flugapparate von paul mac cready” (1981), reprint, *form*, Nr. 199, Nov./Dec. 2004, pp. 53 ff.
**Anonymous:** “Das gelb-weiße Designkonzept ist durchgeflogen,” *Design Report*, no. 9, 1989, pp. 13 ff.
**Anonymous:** “Freudvolle Formen als Ausdruck sachlicher Funktionalität. Wolf Karnagel gestaltete für die Deutsche Lufthansa die neue Bordausstattung,” *form*, no. 116, 1986, pp. 25 ff.
**Anonymous:** “Lufthansa – die konsistente Entwicklung einer Corporate Identity,” in: Birkigt/Stadler/Funck (eds.), *Corporate Identity – Grundlagen, Funktionen, Fallbeispiele*, Frankfurt am Main, 2002, pp. 421 ff.
**Bachinger, Richard:** *Unternehmenskultur – Ein Weg zum Markterfolg*, Frankfurt am Main, 1994.
**Coutts Clay, Jennifer:** “Ways ahead,” in: *Jetliner Cabins*, London, 2003, pp. 177 ff.
**Deutsche Lufthansa AG (ed.):** *Follow me*, Köln, n. d. (1980s).
**Deutsche Lufthansa AG (ed.):** *Wir fliegen wieder nach Berlin*, Cologne, 1990.
**Deutsche Lufthansa AG (ed.):** *Die Zeit im Fluge – Geschichte der Lufthansa seit 1926*, Frankfurt am Main, 2002.
**Edelmann, Klaus Thomas:** “Irrflug?,” *Design Report*, no. 9, 1989, p. 12.
**Eisenbrand, Jochen; Vegesack, Alexander von (ed.):** *Airworld – Design und Architektur für die Flugreise*, exhibition catalaogue, Vitra Design Museum, Weil am Rhein, 2004.
**Esslinger, Hartmut:** “Kranich auf neuem Kurs,” *form spezial*, no. 2, 1998, pp. 58 ff.
**Franck, Bernd:** “Und wie man sich zeigt, so ist man. Rückblick. Vor 35 Jahren verpasste Otl Aicher der Lufthansa ein neues Erscheinungsbild,” *form spezial*, no. 2, 1998, pp. 68 f.
**Geberzahn, W. O., und Grützner, Juliane:** “Mode, Mysterium und Design. Gabriele Strehle formt das Label Strenesse zu einer internationalen Marke,” *Design Report*, no. 10, 2001, pp. 52 ff.
**Harms, Ingeborg:** “Farben des Himmels. Pop-Kraniche, Orient-Formel und Hedonisten-Schick: Die Mode der Flugbegleiter. Eine Stilkritik,” *Frankfurter Allgemeine Sonntagszeitung*, 30. 6. 2002, pp. V2, V3.
**IF International Forum Design GmbH (ed.):** *Designpreis Industrie Forum Hannover 2002*, Hanover, 2002.
**Jens, Tilman:** “Des Kranichs neue Kleider,” *Design Report*, no. 12, 1994, pp. 34 f.
**Jens, Tillman:** “Unter den Wolken: Ein notorischer Vielflieger erlebt das neue Lufthansa Design” *form spezial 2*, 1998, pp. 66 f.
**Knies, Daniel:** “Flugzeuge im Bauch – Essen über den Wolken,” in: Hochschule für Gestaltung Offenbach am Main (ed.): *auch anders – Jahresbericht 2004*, Offenbach am Main, 2004, p. 226.
**Loschek, Ingrid:** *Reclams Mode- und Kostümlexikon*, Stuttgart, 1999.
**Lovegrove, Keith:** *Airline. Identity, Design and Culture*, New York, 2000, pp. 109, 126 f.
**Mies, Petra:** “Mode zum Abheben,” *Frankfurter Rundschau Magazin*, 5. 1. 2002.
**Müller, Birgit:** “Abenteuerlicher Auftritt,” *Design Report*, no. 5, 1997, pp. 39.
**Prinz, Susanne:** *Besteck des 20. Jahrhunderts. Vom Tafelsilber zum Wegwerfartikel*, Munich, 1993, pp. 71 f.
**Rahskopff, Hermann:** Minutes from memory after a visit of Otto Firle in Düsseldorf on November 18, 1965, LH archives, Frankfurt am Main.
**Schmidt, Doris:** “Geschirr in der Luft,” in: *Zeitgemäße Form. Industrial Design international*, Munich, 1967, p. 91.
**Schweer, Michael:** “Die richtige Kurve ist seine Stärke,” *md*, no. 2, 1991, p. 49.
**Schweer, Michael:** “Hightech in Form. Müller/Romca Industriedesign im Porträt,” *md*, no. 7, 2004, pp. 52 ff.
**Steguweit, Christian:** “Corporate Identity – Problem oder Lösung für die Fragen der 90er Jahre – Eine Standortbestimmung am Beispiel der Lufthansa,” in: Norbert Hammer: *Die stillen Designer – Manager des Designs*, Essen, 1994, pp. 103–116.
**Steguweit, Christian:** “Lufthansa. Typologie und Konsequenz einer Corporate Identity,” in: Klaus Schmidt (ed.), *Corporate Identity in Europa: Strategien, Instrumente, Erfolgreiche Beispiele*, Frankfurt am Main / New York, 1994, pp. 135–142.
**Sweet, Fay:** *frog – form follows emotion*, New York, 1999.
**Willich, Dieter:** “Vom Ursprünglichen und Ähnlichen … oder die Quadratur des Plagiarius: Baumann gegen Karnagel, Karnagel gegen Baumann, Baumann gegen Willich,” *form*, no. 130, 1991, pp. 64–67.

Airbus A321-100 im „Parabeldesign“ der 1950er Jahre aus Anlaß des 50. Geburtstags der Lufthansa
Airbus A321-100 with the 1950s “parabola” design on the occasion of Lufthansa’s 50th anniversary

**Bildnachweis**

Florian Aicher, Leutkirch: S. 59 (o)

Frog Design Europe, Herrenberg: S. 74 (u.), 111 (o.)

Walter Hanel, Bergisch Gladbach: S. 15, 17, 18

Hochschule für Gestaltung, Offenbach (Foto: Oli Hege, Frankfurt a. M.): S. 135

Hollin & Radoske, Frankfurt (ehem. „Bhar"): S. 75, 76, 80, 81

Wolf Karnagel, Berlin: S. 114, 127, 128

Müller / Romca Industrial Design, Kiel / Hamburg: S. 102

C. Hugo Pott, Solingen: S. 120

Hans „Nick" Roericht, Ulm: S. 123, 124 (u.)

Schönwald AG, Schönwald: S. 132 (u.)

Topel und Pauser (heute: SquareOne, Düsseldorf, und bluequest Design Group, Darmstadt): S. 132 (o.), 134

Vitra Design Museum, Weil am Rhein (Foto: Andreas Sütterlin): S. 124 (o.)

Alle anderen Abbildungen stammen aus den Bildarchiven der Deutschen Lufthansa AG in Frankfurt am Main und Köln.

Sollten wir einzelne Bildrechte nicht berücksichtigt haben, so bitten wir dafür um Nachsicht. Wir werden dies baldmöglichst korrigieren.

**Photo credits**

Florian Aicher, Leutkirch: p. 59 (t.)

Frog Design Europe, Herrenberg: pp. 74 (b.), 111 (t.)

Walter Hanel, Bergisch Gladbach: pp. 15, 17, 18

Hochschule für Gestaltung, Offenbach (photo: Oli Hege, Frankfurt a. M.): p. 135

Hollin & Radoske, Frankfurt (formerly "bhar"): pp. 75, 76, 80, 81

Wolf Karnagel, Berlin: pp. 114, 127, 128, 129

Müller / Romca Industrial Design, Kiel / Hamburg: p. 102

C. Hugo Pott, Solingen: p. 120

Hans "Nick" Roericht, Ulm: p. 123, 124 (b.)

Schönwald AG, Schönwald: p. 132 (b.)

Topel und Pauser (today: SquareOne, Düsseldorf, and bluequest Design Group, Darmstadt): pp. 132 (t.), 134

Vitra Design Museum, Weil am Rhein (photo: Andreas Sütterlin): p. 124 (t.)

All other illustrations are taken from the photo archives of the Deutsche Lufthansa AG in Frankfurt am Main and Cologne.

If we have inadvertently infringed upon individual photo rights, then we ask for your indulgence. Corrections will be made as soon as possible.